The Telephone Enterprise

The Johns Hopkins / AT&T Series in Telephone History

Louis Galambos
The Johns Hopkins University
Series Editor

The Telephone Enterprise: The Evolution of the Bell System's Horizontal Structure, 1876–1909, by Robert W. Garnet

The Anatomy of a Business Strategy: Bell, Western Electric, and the Origins of the American Telephone Industry, by George David Smith

From Invention to Innovation: Long-Distance Telephone Transmission at the Turn of the Century, by Neil H. Wasserman

Theodore Newton Vail, president of the AT&T Company, 1885–87 and 1907–19. Vail's
second term as president of AT&T was his more important. During this period he reorganized the overall Bell System into an efficient, functionally specialized enterprise; negotiated an accommodation with the government at both the state and federal levels,
which, by and large, mitigated concerns over Bell's monopoly; and strengthened the
finances and the market position of his enterprise. On 15 June 1919 Vail became chairman of the board of AT&T, a post that he held until his death in April 1920.

The Telephone Enterprise:

The Evolution of the Bell System's Horizontal Structure, 1876–1909

Robert W. Garnet

The Johns Hopkins University Press
Baltimore and London

The Johns Hopkins University Press, 701 West 40th Street,
Baltimore, Maryland 21211
The Johns Hopkins Press Ltd, London

The paper in this book is acid-free and meets the guidelines for permanence and durability of the Committee on Production Guidelines for Book Longevity of the Council on Library Resources.

All photographs and documents are reproduced with the permission of AT&T. No further reproduction should be made without permission. AT&T retains all rights to its documents and photographs.

Library of Congress Cataloging in Publication Data

Garnet, Robert W.
　The telephone enterprise.

　(The Johns Hopkins/AT&T series in telephone history)
　Bibliography: p.
　Includes index.
　1. American Telephone and Telegraph Company—History.　2. Telephone—
United States—History.　I. Title.　II. Series.
HE8846.A55G36　1985　　　384.6′065′73　　　　84-43080
ISBN 0-8018-2698-5

To Dorothy and Joseph,
with love

Contents ❧

Editor's Introduction ❧

Twenty-five years ago I was trying to get started on a career as a professional historian, but I was having some difficulty. I had a Ph.D. from a reputable institution. I even had a job. But I had become very interested in the past activities of certain groups of American businessmen, in the organizations they had created, and even the ideas they had; and this left me at cross-purposes with my own profession. At that time there was very little business in American history. There were only a few professional historians who were interested in describing or analyzing business behavior, in its own right, and the work of those who were writing business history was acknowledged but seldom read or seriously discussed by their peers. That bothered the business historians, and because I was a fledgling professional just then joining their ranks, it bothered me too.

Insofar as business played any role in the courses historians taught and the books they wrote, the emphasis was largely on the problems that business had created for Americans and their political system. In this version of the past, rapid industrialization and the rise of the modern corporation had divided America into two camps: liberals, who occupied one of these camps, had studied the problems generated by business and decided that a more active national government should (among other things) curtail business power, regulate business behavior, and actively support those individuals and groups who lacked the power to cope on their own with the business system. On the other side of this battle line stood conservatives who fought tooth and checkbook against liberal political innovations. For most of the most prominent historians of that day, these battles were the centerpieces of history. All history was not past politics, but the most important aspects of it clearly were. Even those historians who were bent on revising this paradigm, the liberal or progressive interpretation of American history, accepted the basic manner in which the primary concerns of the discipline and thus the content of the scholar's version of American history were then defined.

There were specialists who dealt with social and economic history, but even the economic historians were not for the most part concerned with the development of business institutions. They should have been, I thought. But in those years the subdiscipline of economic history was taking a turn toward economic theory, toward the neoclassical model, the analysis of aggregate patterns of economic growth through quantitative studies. In this brand of New Economic History, business decisions and business institutions were either assumed to have been constant over time and space or were left out entirely because they could not be used to explain those fluctuations in Gross National Product that were defined as the crucial phenomena in America's past.

Neither the neoclassical economists nor the political historians paid much heed to the work of the small band of scholars (many of them associated in some way with Harvard University's Graduate School of Business Administration) who had for some years been writing histories of business firms and biographies of business leaders. These academicians had staked out a subdiscipline that they called business history. By the 1960s it had all of the identifying marks of a profession: its own leaders, its academic journal, and its special array of knowledge, concerns, and values. What business history lacked was an audience.

The responsibility for this rested in part with the business historians themselves. All too often they wrote lengthy, ponderous volumes distinguished more by their attention to detail than their concern for questions of broad import. Like the average doctoral dissertation, these tomes told you more about their subject than most readers wanted to learn. The one general problem that received attention had been inherited from liberal historiography and in particular from the work of Matthew Josephson, who had years before characterized America's business moguls of the late nineteenth century as *The Robber Barons*. The name had stuck. Business historians had in the years that followed mined great piles of detail and extracted from them these nuggets of truth: they concluded that the businessmen they studied were in fact industrial statesmen, not robber barons, and the enterprises the businessmen had created contributed to the nation's progress.

Their pleading on this point was, however, largely ignored. They were charged with being too cosy with the businessmen they studied, too inclined to ignore the general public interest. That was why, the critics said, most of their books were unread, even by other specialists. When these studies did on occasion command some attention—as did, for instance, the work of Thomas C. Cochran and that of Alan Nevins—the business historians still could not budge the profession's consensus that the daily stuff of business behavior, corporate administration, and business leadership was somehow peripheral to the history of the United States.

Today that situation has changed to a remarkable degree. There is more real business in American history and more real history in business. In part this transformation can be traced to the emergence of a new public mood in the nation—a mood that has created a more widespread interest in business affairs and that has filtered into the historical profession, subtly reshaping its evaluation of the business system. Since the 1960s most Americans have become more interested in understanding business than in pillorying robber barons. There were, after all, ominous signs in the 1970s that the national economy was in deep trouble. Battered by inflation, energy problems, and fierce competition from abroad, American business appeared to be more in need of active support than government controls. Other nations, especially Japan, gave their businesses more direct assistance than the American government did. As Japan and other countries made deep inroads into U.S. markets, Americans began to reexamine their premises about business-government relations; they sought assurance that their businesses would under the right conditions be able to recapture their lost glory. America's past eras of rapid growth—even the era of the late-nineteenth-century moguls—began to look like a golden age of prosperity.

Meanwhile, modern liberalism had been losing steam as a political ideology. To some extent liberal objectives had been achieved. To a considerable measure, too, Americans had found cause to become cynical about certain liberal goals and past accomplishments. Having created a formidable administrative state, in part along liberal lines, Americans began to question how and in whose interests it operated. The regulatory system, for instance, seemed all too often to function in ways that neither liberals nor their opponents had anticipated. Agencies launched to serve the public interest had been shown to be self-interested in buttressing their own power. The state had reached such size and complexity that coherent, effective leadership seemed almost impossible to achieve in normal peacetime conditions. These perceptions eroded the enthusiasm for liberal ideology. They left many Americans searching for new ways to understand the present and encouraged historians to look for new ways to understand the past.

As scholars reexplored the American experience, they became more and more interested in the manner in which the country's business institutions had evolved. In part this interest reflected changes within the profession of business history, where there has been a shift toward more analytical studies that raise and answer questions of considerable significance to American society. Particularly important in this regard are the works of Alfred D. Chandler, Jr., whose two major publications are *Strategy and Structure: Chapters in the History of the American Industrial Enterprise* (1962) and *The Visible Hand: The Managerial Revolution in American Business* (1977). Both books are internal histories of the business system. Both ignore the old robber baron question. Both build intellectual bridges to other disciplines and to other subdisciplines within history. With the publication of *Strategy and Structure* and *The Visible Hand,* business history became on balance a net exporter of ideas to the rest of the profession. When the latter book won both the Pulitzer Prize and the Bancroft Prize, it aptly symbolized the manner in which the historical profession's perception of business' role in the American past has been transformed.

Historians in almost every corner of the discipline were by this time attentive to the history of business. This was certainly the case among historians of technology. They looked upon technological change as one of the primary forces that had shaped American history, and to understanding technology, they had to understand the business setting in which most of the nation's significant technological advances had been implemented. Business history and the history of technology became in effect partners in the process of reexploring American history.

Social historians also began to make use of the findings of business historians. When social historians peered into nineteenth-century American communities, for instance, they found societies that were centered on and frequently reshaped by changes in their leading businesses. When social historians looked at the country's modern national institutions, they discovered that business had frequently provided the models for, and sometimes the values that permeated, these organizations. The patterns of interaction were complex and at times contradictory. They could certainly not be understood if one left out the business element or simply assumed that businessmen were antagonistic to the changes that were most significant to the society as a whole.

Even political history has been restructured in recent years, and as scholars reexamined the role of business in the polity, they came to appreciate how varied and ambiguous the relationships have been. No longer do they characterize the progressive era of the early 1900s in terms of a simple struggle between liberal reformers and conservative tycoons. Nor are the 1920s or for that matter the 1950s described as mere interludes between surges of liberal reform; important changes took place during these decades and frequently businessmen actually promoted reform or reshaped government administration in significant ways. Business, scholars now recognized, was closely linked to the emerging administrative state in America throughout the century, and business organizations and concepts provided important elements of what came to be called the associational state or the corporate-liberal commonwealth. As historians studied that commonwealth, the manner in which it functioned, and the way it influenced the development of policy and the distribution of power, they gave to business history a central position in our nation's past.

The new work on business-government relations, as well as the analyses of the structure and performance of business organizations, was built on a foundation of monographic studies that could not have been written if there had not been more history in business in recent decades. By this I mean that more companies have recognized the need to save their records in some systematic way and to make them available for serious historical research. More public and private archives have acquired business manuscipts. More firms have sponsored historical studies. More businessmen and companies have also contributed to the endowments that sustain academic appointments in business history and related subjects. Business support, both direct and indirect, has yielded significant intellectual results, consequences that surely could not have been anticipated during that long era when the history of business was on the outside looking in—outside of the major currents in historical scholarship. Now that situation has changed and this result is a tribute to the firms, the universities, the libraries, and the scholars who long before the fruits could be harvested sustained the development of this subdiscipline in the modern historical profession. In recent years, as business history began to flourish and to find its audience, the number of organizations contributing to this effort increased sharply.

It was in this intellectual and academic context that the American Telephone and Telegraph Company decided to strengthen its archival program and to place a new emphasis on its own history. Concerned that much of the evidence upon which studies of telecommunications had been based came from adversarial court proceedings, regulatory commissions, and legislative investigations, the company decided to open its historical records to all bona fide scholars. As part of this effort, AT&T announced in 1984 a graduate research scholarship to support work that would draw upon the business's records. It decided, too, to assist and promote specific historical studies of the company's development over the more than a century that the firm and its predecessors have been in business.

AT&T and the Johns Hopkins University Press, partners in this enterprise, wisely decided to publish a series of volumes by different authors rather than the sort of single-volume company history—a kind of business biography—that had characterized the field in its early years. In a series, various authors could focus on particular analytical themes. They could each develop their own personal evaluations, adding an element of variety that was missing even from the multivolume firm histories when they were written by a team working under the direction of one scholar. Since the Johns Hopkins/AT&T Series is open-ended, the historians involved need not feel compelled to add to their accounts those dreary chapters that result from a misplaced desire for completeness or the scholarly disease of wanting to say the last word about a subject. In this sense the series represents an innovation in business history, one that seems especially appropriate given the size and complexity of the firm and the importance of science and technology in the history of the nation's telecommunications. No single historian is likely to be equipped to handle all of the important facets of AT&T's history. No single perspective is likely to do justice to the variety of histories embodied in the development of the nation's telephone system.

The first volume in this series is Robert Garnet's study of the evolution of the Bell System's horizontal structure* between 1876 and 1909. As Garnet shows—and some readers will be surprised to learn—the telephone enterprise began as a relatively flimsy community of interests based on the orginal Bell patents. Underfinanced and devoid of professional management, the undertaking evolved in the years he examines into one of the first truly modern corporations in the United States. The base of that corporation was provided by AT&T and the Bell Operating Companies. This foundation was created very gradually over a period of more than thirty years. It was shaped by a variety of forces, including the industry's fast-changing technology and its evolving political environment.

In the early years of the telephone—which the author describes with special care—there was actually no Bell system in the modern sense, only a congeries of businesses linked by the slender thread of the patent rights. By the second decade of the twentieth century, however, AT&T was a different type of firm, for that day a hybrid of sorts. It was already one of the largest companies in the United States. By that time the company was far more centralized then it had been before, and yet the different operating companies still exercised substantial discretionary authority. At the heart of Garnet's volume is his analysis of what and who shaped that unique blend of corporate structure and style. As he shows, this was a history in which individuals and their decisions had something to do with

---

*Horizontal structure or integration refers to the combination of units performing the same basic function, in this case the provision of telephone service. It is usually contrasted with vertical integration, which refers to the combination of sequential steps (for instance, the manufacture of equipment and the use of the equipment to provide telephone service) in the business process. The second volume in this series, George David Smith's *The Anatomy of a Business Strategy*, is a study in vertical integration in the Bell System.

what happened. Moreover, the shapers of this history were not always the chief executive officers of the business. Garnet has dug out of the archive and saved for history the contributions of such leaders as E. J. Hall, who emerged as one of AT&T's chief spokesmen for standardization and technological integration in the telephone system.

The type of system that Hall, Theodore Vail, and others perfected came to have carefully articulated goals, a structure well suited to its business strategy, and a deeply rooted corporate culture. It attempted as well to maintain relatively stable relationships with the national and state governments and with its competitors in the telephone industry. The firm's goals, as Vail stated them, were "One system, One policy, Universal service." The Bell System that embodied these objectives promoted innovation, as well as economic stability in a regulated setting; it supported with significant resources a vigorous program of research and development within the company, while seeking and sustaining a leading position in all of its domestic markets. The institution whose origins Garnet carefully analyzes was to endure from the Vail years until 1984. Today, after the national network has been broken up and the American Telephone and Telegraph Company has set forth on a new corporate course, it is especially appropriate that we look back with Robert Garnet to the beginnings of the Bell System.

Louis Galambos

The AT&T Corporate Archives ❧

It is the policy of AT&T to preserve all company records and artifacts that document its history:—the evolution of its structure and organization, the development of its products and services, and the evolution of corporate policies. Therefore, in addition to record retention requirements imposed by law and retention standards adopted by individual departments for their own purposes, the potential corporate historical value of records will be taken into account in the retention and disposal process. The company's central archival organization has responsibility for assessment of the historical importance of all company records and artifacts and the designation, accession, and preservation of archival records.

The company policy is to stimulate scholarly awareness and use of the materials. AT&T has undertaken to make these records and artifacts available for corporate and approved scholarly reference and use through a systematic program of accessioning and cataloguing, by the establishment of adequate reading room facilities, and by direct preparation of publications.

Preface and Acknowledgments ❧

Until 1984 the Bell System was the world's largest corporation. Ubiquitous in the scope of its operation, it nonetheless furnished what many came to regard as a service uniquely personal in dimension. Fully integrated as a business, in its complexity and interdependence the Bell System was virtually unmatched elsewhere in American industry. It was an enterprise driven by technology and shaped by a complicated blend of entrepreneurial vision, commercial ambition, and public policy. And for much of its history it was a monopoly, sometimes regulated, sometimes not.

It is ironic that the events leading up to the dismantling of the Bell System came to rekindle a desire within the corporation for an account of how and for what reasons its various components were put together in the first place. The 1970s witnessed a profound reexamination of the Bell System's place in American society, a debate that took place in the economic literature of the day as well as within the regulatory and judicial arenas. As part of this dialogue, old myths concerning the firm's early development were resurrected, some in defense of the company's prevailing structure, others in criticism of it. Virtually all of them failed to do justice to what was an extraordinarily complex history; taken together, they have established the need for a fresh, scholarly look at the company's past.

This particular study began in 1977, when I was asked by AT&T to examine the impact of regulation on the configuration of Bell operating company boundaries. Later my project was expanded to embrace questions concerning the general horizontal development of the Bell System, including the technical, legal, operational, and administrative relationships that arose between the parent company and its associated telephone businesses during the early and middle years of the firm's growth. With the full support of the AT&T

management, I was encouraged to expand the scope of my study, to turn it into a broadly conceived history of the enterprise. It thus became my dissertation, written for the Department of History at The Johns Hopkins University; and from that, this book.

In preparing this manuscript for publication, I received valuable counsel and guidance from several quarters. Within the AT&T Company, Philip Haff and Robert Lewis opened doors and furnished the encouragement and direction without which this study could not have been completed. Much of the credit for securing the funding for this project during its formative stages belongs to Phil Haff, whose sense of history and of its importance in understanding the business is rivaled only by his sense of humor. That Robert Lewis, his successor, continues to manage AT&T's archives and historical publication program bodes well for those scholars who will conduct further research into the company's past.

Also worthy of mention are those within the AT&T Company not directly associated with this project who gave this manuscript the time and attention that it needed before it went to press. Richard Priest, AT&T's recently retired archivist; Alvin von Auw, former vice president and assistant to the chairman, whose reflections on more contemporary events affecting the enterprise are recounted in his own book; and George Schindler, of Bell Laboratories, are among those that come to mind. The manuscript improved immeasurably as a result of their assistance and counsel.

While a consultant with the Cambridge Research Institute I had the benefit of advice and criticism from several colleagues, many of whom conducted their own research into the Bell System's historical development. Laurence Steadman, who studied the financial affairs of the business, and William S. Quinn, who examined the early development of the long-distance business, were among the many upon whose work I depended in shaping my own ideas regarding the company's evolution. David Kiser, Fred Cardin, Neal Wasserman, and Nancy Needham were among the others. I am especially indebted to George Smith, now of Winthrop Associates, and Alan Gardner. Smith, whose own work on the acquisition of Western Electric complements my own and is part of the series to which this study belongs, has been a thoughtful and valuable critic. Alan Gardner has been an invaluable source of advice on a whole range of subjects, most particularly on the technical aspects of the Bell System's evolution.

I am indebted as well to two prominent business historians. It was for James P. Baughman, who has since left Harvard to take up residence at the General Electric Company, that much of my earlier research was done. Professor Baughman was at that time preparing an overview of the Bell System's technological and organizational development for testimony in *U.S. v. AT&T*. Louis Galambos, under whose direction this study was continued, bolstered my enthusiasm for the project at the moments when it sank, sharpened my ideas and arguments when they lost their focus, and helped me uncover relationships that were not always obvious to me, being so close to the subject of study. Since the days when I was enrolled in his undergraduate course in American economic history at Johns Hopkins he has inspired my interest in the history of business and an appreciation for its role in our nation's development. And as many of his students surely acknowledge, Galambos's hand is one of a tolerant but precise editor and teacher.

I have also enjoyed the benefit of advice from William Freehling, Carl Christ, William Leslie, and Francis Rourke, of The Johns Hopkins University, who as members of my dissertation committee read an earlier version of this manuscript. Thanks also to Debbie DeFago, Maria Fernandez, Linda Lee, and Sharon Widomski, who suffered in good humor through what must have seemed an unending train of revisions.

Ultimately, the opinions and conclusions expressed in this manuscript are mine, not AT&T's. I must bear responsibility for whatever mistakes in interpretation or fact remain and for the ideas presented. But with the support and counsel of so many, one surely approaches this responsibility with confidence. My respect for the historical profession, and for the process of historical writing and investigation, leads me to believe that such errors will not long remain a part of the permanent record. My hope is that this manuscript will interest others in examining a firm that has been crucial to the economic and social development of our nation and that they will find that my book provides a useful perspective on AT&T's remarkable past.

The Telephone Enterprise

CHAPTER 1 🐦

The Great Yankee Invention

IT WAS FITTING THAT THE YEAR Alexander Graham Bell invented the telephone, Americans proudly observed their nation's centennial in a grand parade of pomp and celebration. What better theme could there have been for this occasion than a century of progress; what better way to erase the bitter memory of civil strife than to reaffirm the nation's faith in the future? In ceremonies repeated across the land, the achievements of the past and of the present were used to christen the promise of the future.

As the site of the nation's Centennial Exhibition, Philadelphia was the center of much of this excitement. Where America's founding fathers once fashioned their political ideals and aspirations into a declaration of faith and independence, city fathers now commemorated the birth of a democracy with a magnificent display of Yankee ingenuity. They invited attendees to explore with them the new frontiers of science and industry, the startling discoveries associated with practical applications of steam power and electricity. There, men such as Alexander Graham Bell could offer an eager public a peek into the future, an image of prosperity and industrial growth founded on the ability to unlock the secrets of new technologies and sciences. Material progress was celebrated, and its creators and promulgators were proclaimed the heroes of the age.

Bell's demonstration of electrical communication delighted those visiting Philadelphia's exhibition. By any standard, the telephone was among the most extraordinary innovations of an age teeming with technological surprises. Bell's demonstration inspired awe and wonderment. There were, no doubt, many who remained certain that the telephone would never eclipse the widely used printing telegraph instrument as a means of communication, but most of the publicists and pundits who speculated about the telephone's future were afire with optimism. The inventor himself cast before the public a prophetic vision of households and businesses across the nation linked together by telephone. The possibilities seemed endless. Its value seemed immeasurable. In a land so vast in territory and so much in need of better communications the telephone was "the great Yankee invention."

Ironically, the modern technology celebrated in the nation's centennial would shortly help usher in a new chapter of American industrial development, one differing greatly in organization and scale from the kind so enthusiastically celebrated in 1876. Bell's success in the laboratory stemmed from a mixture of hard work, good fortune, and imagination; his accomplishments symbolized the triumph of individual effort. The enterprise that he helped launch, however, would become a collective effort of such size and wealth that it would periodically generate profound uneasiness in the larger society that it served. The Bell System and the nation's other corporate undertakings would force Americans to change many traditional patterns of behavior and to reevaluate their traditional values. They would breed conflict and a new political movement; they would make Americans grapple with new questions about the exercise of power and the distribution of wealth in their society.

Many who witnessed the beginnings of the modern corporation felt uneasy and distrustful of the motives of the founding entrepreneurs; in time they became apprehensive about the impact that such institutions would have on the American experience. Frequently, however, their apprehensions were blended with an admiration for corporate efficiency and the capacity of big business to organize production and create wealth on an unprecedented scale. Whatever their response, most Americans seem to have agreed that the modern corporation played a central role in their society's development, and it is therefore not surprising that scholars have paid considerable attention to the sometimes painful adjustment the nation made to the rise of the giant firm. Indeed, several generations of scholars have made the emergence of big business a central phenomenon in their historical interpretations of late-nineteenth- and early twentieth-century America.[1]

As one of the charter members of the country's growing family of large, integrated enterprises, the Bell System was involved in the contemporary political and economic controversies surrounding the rise of big business. Moreover, in subsequent years it has continued to receive a substantial amount of attention. Add to this the ubiquitous position that the American Telephone and Telegraph Company has occupied in the communications industry throughout most of its one-hundred-year history, the great research facilities that it has developed in the twentieth century, and the large number of Americans who have owned shares in one or more of the Bell enterprises, and it becomes clear why the history of the Bell System has attracted so much attention. Its role in history went far beyond the economic realm. Bell's Yankee invention had an impact on the American pattern of social and personal development, on urban and then rural life, on public as well as private institutions.[2] Although the telephone network in the United States was essentially a private, capitalistic undertaking, the Bell System came under government regulation early on, and its history came to blend with the history of the regulatory movement in the states and the nation.

Over the years, the history of the Bell System has accumulated its share of myths, replete with scoundrels and heroes. The inventor of the telephone was worthy of Hollywood's attention. Alexander Graham Bell harbored a lifelong passion to understand the mechanics of human communication—a passion that played no small role in his subsequent discoveries. But he actually stumbled on the general principles governing his remarkable device by accident. Moreover, he derived little satisfaction from the business he founded. Shortly after the crucial invention was patented and its acceptance as a useful means of communication assured, he returned to his other endeavors. He seemed content to bask in the worldwide recognition that he received as an inventor, and he left to businessmen the task of building and managing the new enterprise.

The challenges facing these businessmen were great indeed. Their first years were spent pushing for expansion while defending the company against patent infringement, a battle that took on new and threatening proportions after the goliath Western Union entered the field. From these years of turmoil emerged the firm's first heroes: Thomas Watson, Bell's tireless, devoted associate, a man whose never-ending search for ways to improve the original apparatus kept the firm in the forefront of a new, rapidly developing technology; Theodore Vail, whose driving personality, administrative skills, and daring business strategies held the outfit together during its most difficult early years; Gardiner Hubbard, the promoter whose vision had a decisive impact on the nature of the company; and the more conservative Thomas Sanders, who, despite the grave uncertainties of the undertaking, risked most of his fortune to get the company on its feet.

These businessmen and Bell himself have attracted the attention of biographers, and their works provide a measure of information on, and insights into, the company's early history. Even those studies that achieved no more depth than newspaper accounts, such as Catherine MacKenzie's biography of Alexander Graham Bell and Albert Bigelow Paine's biography of Theodore Vail, can be useful.[3] More recently, Robert V. Bruce has reexamined Bell's life in a scholarly biography that draws upon a collection of Bell papers now at the Library of Congress. Bruce captures all of the tension and excitement of the early years of the telephone, and he examines in admirable detail the complex litigation that took place over the important telephone patents.[4]

To the historian of the firm, these studies—and others, such as the biography of William H. Forbes, who headed the company's second generation of leaders[5]—are both useful and frustrating. They are useful in part because they help the historian visualize events and personalities as their contemporaries saw them; moreover, they advance the intriguing idea that men who made specific decisions at specific times in the past shaped the development of this business and others like it. But the biographies are ultimately frustrating because their subjects did not devote all of their career to this single business (as do most modern, professional businessmen), and the authors of these books can seldom delve very deeply into the evolution of the business itself.

Fortunately, there are also in print some business histories of the Bell System, histories that focus primarily on the firm, not its leaders. One of these is J. Warren Stehman's laudable, scholarly account *The Financial History of the American Telephone and Telegraph Company*.[6] Stehman, an economist at the University of Minnesota, provided his readers with a cogent analysis of the Bell System's complex financial activities. While he did not try to delineate the internal structure of the firm, he traced in some detail the changes that took place in the financial relationships between the parent company and its licensees. He was, moreover, more interested in long-term strategic and financial trends than he was in the tenure of particular corporate administrations. There were, he suggested, three stages in Bell's history: a brief competitive period; fifteen years of patent monopoly (1880–93); and a period characterized by intense competition, from 1894 to the early 1920s. He gently observed that "most industrial organizations formulate their policies and conduct their business in a somewhat different manner under competition than under monopoly." The Bell System, he said, had over the long run performed well in terms of service to the public. Profits had not been unreasonably large, and the national system was integrated in a manner that gave the American people "the best and most efficient telephone service." It was true, Stehman concluded, that during the period of patent monopoly "the attitude assumed toward the public by many of the officials of the various Bell companies appears to have been arrogant and discourteous." But the Bell System had reformed. "During the past decade and a half," he said, company "officials have seemed to desire sincerely to work in harmony with public authority and to perform, to the best of their ability, real public service."[7]

Stehman established a chronology that even today is adopted by many who describe the Bell System's development, but his favorable evaluation of AT&T's record of performing "real public service" has been challenged on several fronts. In the depression-ridden 1930s many Americans became hostile toward the brand of big capitalism that seemed to have caused the country's economic woes. In the latter part of the decade, the administration of Franklin D. Roosevelt focused these public hostilities by launching both an antitrust drive and a major investigation of the large corporation's impact on the economy.[8] Meanwhile, the Federal Communications Commission (FCC) began conducting its own investigation of the business practices and organization of the Bell System.[9] The latter inquiry left a legacy of over sixty volumes of transcripts, produced more than seventy staff reports and over a thousand exhibits, and sparked an extraordinary controversy within the agency itself over the conclusions to be drawn from the probe.

The FCC investigation also prompted two authors to write histories of the Bell System: Horace Coon's *American Tel & Tel: The Story of a Great Monopoly* and N. R. Danielian's *AT&T: The Story of Industrial Conquest*. Both were extremely critical of their subject, and both were written in the tradition established by Matthew Josephson's famous study *The Robber Barons*.[10] Of the two books, Danielian's seems to have attracted the most attention, in part because of the vigor with which he attacked those concentrations of economic and political power that AT&T symbolized. In his view, AT&T's management was, by 1907, little more than a "servant of the banking interests," and the bankers part of an intricate web of interlocking directorates binding the country's industrial and financial forces together in a disciplined, aggressive oligarchy "exhibiting all the economic and political propensities of a national state in its most imperialistic moods."[11] The holding company structure, a mechanism for financial control, became the key to understanding the Bell System (and by implication the overall American economy).

Danielian covered many aspects of Bell's history, but the one that aroused his passion was the structure of corporate control at AT&T. In effect, he turned Stehman's institutional analysis on its head. Where Stehman saw in organizational growth an organic process constrained by market and financial considerations, Danielian described a virtually unhindered consolidation of economic and political authority. Where Stehman saw a "natural tendency in the telephone industry" to increase the size of the operating unit, Danielian recognized only a concerted effort to eliminate the position of minority shareholders in merging exchange systems. Where Stehman represented Theodore Vail as a forward-looking industrial captain, Danielian saw him as nothing more than a dedicated soldier for the Morgan interests.[12] Both authors measured AT&T against a model that stressed the inefficiency and dangers of unfettered monopoly, but Danielian's ideological context involved as well a critique of the entire laissez-faire system of political economy in the United States.

With many of his contemporaries Danielian shared a vivid sense that America had reached a crossroad. Down one path the nation would preserve its democratic polity and traditional form of competitive capitalism. Down the other path was corporate America, a new system that aroused Danielian's deepest fears. He was concerned about the marriage of colossal financial interests with industrial power and the rise of managerial capitalism. The problem was not AT&T in particular, although the Bell companies were surely no small part of this disturbing pattern. Actually, Danielian conceded that he had found "no scarlet patches reflecting on [AT&T's] financial chastity—no financial double dealing by officers, no exhortation by insiders, no juggling of markets for the benefit of a favored few. The System still remained pure before the public eye—a shining example of peaceful coexistence under government regulation—a good monopoly." Coon agreed and characterized the firm as "the finest example of capitalism."[13] Both authors were perturbed, however, that the power flowing from such economic success threatened to seep into the political arena, tilting the delicate balance in favor of a small group of conservative forces. Danielian warned, "Whether the United States shall continue a democracy or shall revert to a form of oligarchic control imposed by corporate states through assumption of state powers today hangs in the balance."[14] Both Danielian and Coon obviously were measuring AT&T against the alarming political events taking place in Europe during the thirties. Events in the United States, they hoped, would follow a different path.[15]

Indeed, the United States did set its own, unique course, but events since that time have not made large enterprise of less interest to either American scholars or the general public. To the contrary, scholarly interest in the history of giant corporations such as AT&T— for many decades the largest corporation in the world—has grown in the years since World War II. Recently, the scholarship of Alfred D. Chandler, Jr., has focused substantial attention on the rise of big business and has shifted our historical perceptions of this phenomenon. All of the authors that I have discussed thus far used a similar approach to the history of the firm—in this case AT&T. They looked primarily at the company's competitive environment, the environment outside the firm, and they adopted a chronology that emphasized two or possibly three sharp breaks in the company's development: the noncompetitive era through 1893 was followed by a long phase of competition. For Danielian, this stage could be broken at 1907, when AT&T's new management adopted an aggressive policy of acquiring competitors. Essentially, however, Danielian, Coon, and Stehman adopted similar positions on the Bell System's evolution, and the same could be said of John Brooks, author of a more recent history published to mark the Bell System's centennial.[16] While he was far more favorable toward AT&T than was Danielian, Brooks looked outside the business for the guideposts to its historical development.

This was not the case with Chandler, whose publications have decisively altered the field of business history. In *Strategy and Structure: Chapters in the History of Industrial Enterprise* and, more recently, *The Visible Hand: The Managerial Revolution in American Business*, Chandler directed attention to the dynamic interaction between the strategic goals and the administrative structure of the large firm in this country.[17] The external factors of greatest impact on business were the available markets and, above all, technology. Technology was of overriding importance in shaping the highly centralized, functionally departmentalized, vertically integrated firms of the late nineteenth century. In Chandler's version of business history an internal process of problem solving, of administrative rationalization, replaced the drive to monopolize as the engine of change. Political considerations of the sort that concerned Danielian and Coon were dismissed as relatively unimportant. The emerging hierarchies of corporate control no longer constituted a disturbing concentration of power; instead, they were a product of management's drive to organize their growing businesses along lines more efficient than those that had existed before the rise of big business.

Although Chandler did not devote much attention to the Bell System, he left little doubt as to the main force shaping its development. The telephone, like the railroad and the telegraph, came to be dominated by large enterprise for technological reasons.

Chandler explained: "As the telephone network began to expand in the 1890s, the pioneering group—the Bell interests—maintained its control of the industry 'through traffic' by means of the American Telephone and Telegraph Company, which built and operated through or long-distance facilities. In modern communications, as in modern transportation, the requirements of high-volume, high-speed operations brought the large-scale managerial enterprise and with it oligopoly and monopoly." It was, he said, "the operational requirements of the new technology" that "brought, indeed demanded, the creation of modern managerially operated business enterprises."[18]

Chandler's impressive synthesis of business history is, of course, attractive to the historian of the Bell System. It focuses on a single, major causal force. It places little emphasis on individuals and their choices, thus leaving the history relatively uncluttered. It avoids the broad sociopolitical questions asked by writers such as Brooks and Danielian. In fact, it leaves out political considerations almost entirely and concentrates on developments within the firm.

In my study of the emergence of the Bell System's horizontal structure, I follow Chandler's lead and look carefully at administrative developments within the system.[19] Particularly alert to those kinds of incremental changes that usually characterize administrations, I examine the extent to which those changes appear to have been a product of technological needs. Moreover, I compare AT&T with the highly centralized, vertically integrated firms that Chandler describes.

In other regards, however, I adopt a somewhat different context from the one used by Chandler. Those political factors that appear to have had a significant impact on AT&T's history are not overlooked.[20] Nor does this story start with the assumption that individuals and their highly particular choices were ipso facto less important than environmental forces in shaping the company's evolution. To the contrary, I look closely at the manner in which the company's leaders anticipated, integrated, and responded to changes inside and outside the telephone system. Finally, I bear in mind that an interest in bringing competition under control was not necessarily inconsistent with an interest in remaining technologically innovative and administratively efficient.

In order to do justice to these several themes and to untangle the various factors that helped shape the Bell System's horizontal structure, we must return again to the very beginnings of the enterprise. Then, all that Alexander Graham Bell had was a novel invention which he had patented and the financial support of two New England businessmen—one of whom was soon to become his father-in-law. In the year 1876 there was no system. At that time there was no corporation, no company as we understand it.

CHAPTER 2 🌱

Humble Origins

FOR THE PIONEERS OF TELEPHONY, there was a wide gap between invention and commercial success. Indeed, much skepticism concerning the practical applications of telephony greeted Bell and his two benefactors as they geared up to turn their patented "liquid transmitter" into a successful business.[1] Much of what has been written about their early struggle to gain commercial acceptance for the telephone is undoubtedly exaggerated, part of the corporate mythology that has come to envelop the firm's history since its founding. But much of it is true. Officials of the Western Union Company, who one presumes should have known better, dismissed out-of-hand Hubbard's offer to sell them the patent rights to Bell's invention for one hundred thousand dollars.[2] The market for this invention was unproven, the telephone itself a mere scientific curiosity.[3]

The early opinions of Western Union leaders were shaped by a relatively narrow conception of just where the telephone fit into the range of services then available to business customers. Bell's device was taken to be a substitute (and an unreliable one at that) for the Morse key instruments then used on the telegraph company's extensive wire network. For the business community, which depended heavily on the written record as a basis for commercial transaction, the telephone seemed to offer few tangible advantages over the telegraph messages of the day. Even among Bell's admirers "so strong was the prejudice and distrust against the telephone" that most gave it less than an equal chance against such well-established means of communications as the printing telegraphs.[4] At best, it seemed, the telephone would round out the existing system of wire communications.

It is not surprising that Alexander Graham Bell was of a decidedly different opinion regarding the value of his invention. From the outset, he confidently predicted that in time "a telephone in every house would be considered indispensable"; that business-men would one day have "no more difficulty in talking with . . . agent[s] a hundred miles away" than they did addressing servants directly over their own residential "speaking tubes." In the spring of 1877 he took to the field to prove his contention. With Frederick A. Gower he toured New England, demonstrating the potential of his new technology. Bell and Gower developed provocative visions of a future "central office" that would link together in electrical con-versation entire communities. Before an assembly of citizens in Hartford, Connecticut, Gower painted a glowing portrait of their city's future in the age of telephony.[5] Local dailies covering these demonstrations gave their readers grand images of a nation wired together by iron circuits; in humor, they reflected on the "awful and irresponsible power" that the telephone would confer upon the average mother-in-law.[6] By the time Bell and Gower had completed their first journey into the field, they had generated a considerable amount of publicity for the invention, and a small market had begun to emerge.

Back in Boston, Bell's two partners—who had helped finance his early work on the harmonic telegraph and its successor, the liquid transmitter—were busy organizing the business. On 9 July 1877 Gardiner G. Hubbard and Thomas Sanders formed the Bell Tele-phone Company of Massachusetts, an unincorporated, voluntary association not unlike the many small commercial partnerships common to New England at the time.[7] Rights to Bell's invention originally entrusted to the patent association were reassigned to Hubbard, the new company's trustee and principal officer. Sanders, whose financial stake in the business had grown larger than those of his colleagues, assumed the post of treasurer, a position from which he could exercise control over disbursements, maintain accounts, and review the contracts negotiated by the firm's trustee. Thomas Watson, Bell's assistant at the time of his early experiments in

Gardiner G. Hubbard, first trustee of the Bell Telephone Company and one of the enterprise's three founders. During his brief stint as Bell's principal executive, Hubbard was responsible for initiating the longstanding policies of leasing instruments and licensing local agents. Though Hubbard was, by and large, removed from a position of authority over company affairs by the group of Boston financiers who helped organize and underwrite the National Bell Company, he remained an influential voice on the company's board of directors until the late 1890s.

Alexander Graham Bell. Although Bell would become the first electrician for the company that bore his name, he would not play a major role in organizing the business. Instead, he chose to bask in the limelight of his scientific achievements and continued to pursue his research into the mechanics of learning and communication.

Thomas Sanders, along with Gardiner Hubbard and Alexander Graham Bell one of the founding partners of the Bell enterprise. Sanders became the company's first treasurer and along with his friends in the Boston financial community was responsible for strengthening the firm's shaky financial condition during the early years of the business. His financial conservatism frequently put him at odds with Hubbard. Although his influence over Bell affairs waned considerably after he brought in a new group of investors to help underwrite expansion in 1879, he remained on the parent company's board of directors until mid-1911.

13

telephony, became the firm's first superintendent of telephone instrument production, and Bell himself assumed the position of company electrician. A five-member board of managers comprising all of the firm's principal shareholders was appointed to establish corporate regulations and policy.[8] The five thousand shares of stock issued to this small circle of officers, employees, and relatives carried no intrinsic value but instead represented a future claim against the commercial value of the patents.[9]

The Bell Company of 1877 was more like a family partnership than a modern business enterprise. Both Gardiner Hubbard and Thomas Sanders had known Alexander Graham Bell for years, first as the teacher they had hired to instruct their deaf children and later as a friend whose curious endeavors in harmonic telegraphy they had modestly underwritten. Two days after the Bell Company's organization, Hubbard's personal ties to the inventor acquired a new dimension. On 11 July, Bell and Mabel Hubbard were married, declaring their vows in storybook fashion during a ceremony at Hubbard's home on Brattle Street in Cambridge.[10] Gardiner Hubbard's brother Charles was a clerk in the firm and a minor shareholder. Watson, of course, had been Bell's close companion for the better part of two years. All of these officers and employees of the company had known each other personally for some time, and their family and commercial links would influence the nature of their working relationships during the formative stages of the business's development.

Hubbard seems to have made many of the most crucial business decisions at the outset, including the decision to lease rather than sell Bell's telephones. Unable to excite the interest of large investors in the commercial prospects of telephony after an exhausting half-year odyssey that took him from the industrial centers of New England to the oil fields of Pennsylvania, Hubbard began enlisting a small army of local agents across the country. These agents were contracted to lease the company's equipment to customers in return for an annual commission, usually 40 percent on each telephone during the agent's first year of operation and 20 percent thereafter.[11]

The first telephone advertisements listed the "terms for leasing two telephones for social purposes, connecting a dwelling house with any other building," as twenty dollars per year; forty dollars per year for "business purposes," payable in advance.[12] These differences reflected Hubbard's conviction that telephone service had far greater value in commerce than it did as an instrument of casual conversation. The strategy of leasing and the prices charged, moreover, were designed with the specific intention of furnishing the company with a steady flow of income with which to build and expand its operations, income that neither Hubbard nor Sanders was able to provide owing to limited personal savings and other financial commitments.[13] The company's leasing arrangements also helped surmount a serious and continuing shortage of capital by shifting onto the shoulders of the local agents the costs of promotion in the field, including all expenditures associated with the marketing, construction, and operations of telephone plants.

But what is perhaps most important, the combined strategy of leasing and licensing enabled the company to cover a market virtually national in scope while maintaining absolute control over the distribution and use of its valuable patented instruments. Despite its innovative technology, Bell's telephone was a relatively simple device, easily imitated. Indeed, between 1878 and 1894 a flood of imitations would pour into the marketplace.[14] The Bell Company was, however, determined and well prepared to defend its patented technology. Detailed leasing records prepared by local agents aided in identifying the existence of bogus instruments in the field and ultimately prosecuting over six hundred infringement suits.[15] With few employees, little capital, and only a slight technological edge over his competitors, real and potential, Hubbard understood from the beginning that the defense of his company's patents would be crucial to its success. He formulated his policies accordingly.

In the beginning, the basic configuration of telephone operations was very different from the modern system of voice communications that we enjoy today. The idea of a switched network was little more than a prophetic vision in 1877. At that time telephony was a relatively simple point-to-point service, usually between one's residence and place of business. Essentially it was what today would be called a private-line operation.

The point-to-point configuration had important ramifications for the way the Bell Company was organized and financed on the local level. Because each private line represented a discrete operation, local agents were encouraged to sell these lines to sub-scribers at a price of between $100 and $165 per mile per circuit.[16] This lightened considerably the burden of underwriting the de-velopment of early telephone facilities for the licensees. It also provided agents with two sources of income: the commission that the Bell Company granted on leased instruments and the profits earned on the construction and sale of the private-line telephone plant. When such properties could not be sold, the start-up capital re-quirements of the business were much higher, and the agent's investment had to be recovered over a much longer term through annual service and maintenance charges. In either case, however, the burden of these expenditures fell on the local licensees, not on the Bell Company.

Hubbard issued the first such licenses to local agents in early June 1877. There was no consistent pattern to these early as-signments, with agency boundaries varying greatly in size and in some cases overlapping. By midsummer Hubbard had three agents working side by side in Boston: one was licensed to build private lines; the other two rented instruments and, where necessary, operated the private systems. In the Midwest and the South, by contrast, large blocks of sparsely settled territory were granted to a relatively small number of licensees, who in turn were encouraged to sublicense sections of their respective franchises to other local investors interested in establishing a telephone business. By the end of his first month as trustee, Hubbard had negotiated agreements for the major cities of Albany and New York; had concluded temporary arrangements for the rest of New England with Bell's partner on the lecture circuit, Frederick Gower; had found agents for western New York State, Ohio, Indiana, and the oil regions of Pennsylvania; and had completed terms for South Carolina, Georgia, and the northern sections of Florida.[17]

Many of these agents had already had business dealings or enjoyed relationships with the firm's founders. The partnership of Stearns and George, which received a limited franchise for portions of metropolitan Boston, had constructed lines for Alexander Gra-ham Bell's earliest transmission experiments outside of his Milk Street laboratory.[18] George Coy, of New Haven, had built temporary facilities for Bell's memorable Connecticut lectures.[19] A license for western Pennsylvania was awarded to John Ponton, a local news-paper publisher who had personal ties to Alexander Graham Bell, and the rights to develop California went to Hubbard's long-time friend George Ladd. As with most of the mercantile ventures of the nineteenth century—and previous centuries as well—personal ties were more important than administrative controls. Where distances were great and communications still poor, personal relationships laid a foundation of trust that bound the enterprise together into a cooperative partnership.

The technical nature of the telephone business, however, ultimately favored agents experienced either in the manufacture and supply of electrical equipment or in the operation of the telegraph industry. Among the company's most successful franchises were those managed by men like Davis and Watts, the owners of a Baltimore electrical shop, and E. T. Holmes, the founder of Boston's largest burglar alarm business. Charles Haskins, of Wisconsin; G. W. Stockley, of Cleveland; William Bofinger, of New Orleans; C. H. Sewall, of Albany, New York; and H. R. Rhodes, a superintendent of lines for the Pennsylvania Railroad, all were experienced tele-graph men who became agents and licensees of the Bell Company.[20] Most of them were considered to be judicious businessmen in their own communities, which helped them dispel the skepticism that remained over the commercial prospects of telephony. Almost all of them appeared to wield enough influence over local affairs to secure the rights of way needed to construct poles and string lines on public properties.

By bringing men such as these into the business, Hubbard was able, without much capital outlay, to expand the company's base of human and financial resources. By the fall of 1877 his cautious early experiments with licensing had given way to a virtual race to appoint agents across the country. Meanwhile, manufacturing arrangements had been concluded with Charles Williams's electrical shop in Boston, and under Watson's supervision, over three thousand in-struments had been tested and shipped into the field by the end of October.[21] Hubbard's pricing policies were designed to enable the firm to tap both the business and residential markets. By guaran-teeing the repair of faulty or broken instruments "free of expense" except in instances of "great carelessness," he assured his customers that they would not have to worry about these new, poorly under-stood devices.[22]

A vigorous and successful promoter, Hubbard was to prove to be somewhat less effective as an administrator. He was more concerned with expansion than with the details of licensing contracts, more worried about increasing output than about how much it would cost. Easily tired by the tedium of managing the daily affairs of the business, he tended to ignore, or at least to discount, the financial strains on his small enterprise. He seemed instinctively confident that net rental earnings would sustain his company's growth. The meticulous Thomas Sanders, on the other hand, was considerably less optimistic. His own records and forecasts indicated that the business was headed for trouble. But for a time he forgave Hubbard his weaknesses as an administrator and negotiated for the business a new, temporary line of credit. He also used his own money to cover the firm's mounting debts.[23]

The shortages in the company's treasury could not, however, be long met with remedial measures. By December, Sanders saw that the enterprise was again on the edge of financial collapse, and he urged Hubbard to consider refinancing their venture with the aid of friends in the Boston investment community. But Hubbard resisted. Determined to avoid any dilution of his control over the business, he viewed the new partnership proposed by his colleague as the first step in an irreversible process that would eventually undermine his authority.

Rumors that Western Union was becoming interested in the telephone business, however, forced Hubbard to temper his opposition on this matter.[24] That the financially undernourished Bell Company might awaken one day to discover itself in competition with the nation's largest, most influential enterprise was possibility enough to shake the confidence of even the most determined and optimistic telephone advocate. Though the Bell interests enjoyed what the company's founders believed to be a strong patent position, given the limited resources at their command, they could little afford to undertake a prolonged and costly legal battle in their defense. Even Hubbard, inclined though he was to minimize such risks, finally acknowledged the hazards of completely ignoring his firm's financial situation. Coming around to Sanders's position, he reluctantly endorsed a compromise that would establish a new and separate venture for the New England territory with "capital sufficient to construct, maintain and lease telephones, call bells and also lines of telephones to parties . . . unable to construct them and own them for themselves."[25]

The organization of a separate company with broad powers to develop the market for telephone instruments and private-line services in New England was a plan that generally appealed to the investors that Sanders had lined up. What they did not greet with enthusiasm was Hubbard's insistence that the new venture assume the status of a subsidiary of the Bell Company. George L. Bradley, spokesman and negotiator for the New England financiers, refused to subordinate his interests or those of his colleagues to Hubbard's authority, a position that brought angry accusations from Hubbard to the effect that the newcomers were planning to administer the new enterprise "independently of the general interests of the Bell Company."[26] In an effort to retain control over affairs in New England, Hubbard tried to enlist the support of Sanders and Watson; but the Bell Company's board of managers rejected this proposal, refusing to allow such "technical difficulties" to stand in the way of reorganization.[27] Isolated in his opposition, Hubbard reluctantly acquiesced. Later he wrote mournfully to Sanders about his ambition to make the original Bell Telephone Company "a power in the land," a venture in which the founders could "all take pride." But now, with the company's "right arm" sacrificed on the altar of financial necessity, the chances of fulfilling his objective, he said, had been compromised.[28]

There was, of course, some consolation for his defeat. Hubbard was elected president of the new enterprise, which would now begin marketing telephones in the New England territory. But this concession barely cloaked the fact that Bradley, the firm's new general manager, had assumed the executive functions once assigned to Hubbard as the company's trustee. Hubbard's authority (but not Sanders's) had been curtailed in return for new capital. The New England Company opened for business on 12 February 1878 with fifty thousand dollars in its treasury and about eight hundred telephones in the field, all of them previously installed by licensed agents.[29] George Bradley, his brother Charles, and two Boston financiers, William G. Saltonstall and George Z. Silsbee, furnished the funds needed to start the business and in return received one hundred thousand dollars in New England Telephone Company stock.[30] Stock worth another hundred thousand dollars had been given to the original patentees for the exclusive rights to market their device in Massachusetts, Vermont, New Hampshire, Maine, and Rhode Island.

The prospects for the business, despite some concern over Western Union's intentions, appeared quite favorable. Hubbard, ever the optimist, anticipated adding another twelve hundred instruments to the company's revenue base in the next fifteen to eighteen months. He encouraged Bradley to diversify his sources of corporate earnings through the sale or lease of call bells. At an average annual rental of $7.74 per telephone, $5.00 per bell, and $35.00 per line, his projections of gross receipts for 1878 amounted to about $35,000 —a respectable return. Operational expenses were crudely calculated at around 40 percent of the firm's yearly income, capital requirements at $52,000. It did not take a sophisticated accountant to realize that such figures, if accurate, would result in a handsome profit, even at the end of the first year and a half of operations. Hubbard himself predicted a profit slightly above $20,000—a return sufficient to earn 10 percent on the company's outstanding stock and much more on the actual cash investment.[31]

Expectations for the New England market were based on the brief experience that Bell officials had had in private-line services. The price paid by subscribers for such unshared, point-to-point facilities was relatively high. Only businessmen desiring a personal, direct communications link between their commercial establishment or factory and their private residence appeared willing to pay the expense of subscribing to Bell's private-line service. Few residential customers saw much advantage in being linked to only one other residence. Until advances in technology dramatically changed the dimension of the Bell Company's market, its prospects were good but not outstanding.

Just such a breakthrough came in early 1878 in the form of the first group of switched central office systems. The switchboard, which became the technological centerpiece of exchange operations, transformed telephony by placing each telephone subscriber in communication with any other subscriber connected to the same central exchange. Greatly enhanced was the value of this service to business and residential customers alike. Exchange technology created a new and far larger market for telephone instruments and services, a market whose exploding demand the Bell interests would find difficult to satisfy.

The exchange was an innovation fashioned by many hands. The commercial prospects of multipoint services were first explored by E. T. Holmes when in May 1877 he connected four banks and the Williams shop over circuits belonging to his Boston-based security alarm business.[32] Holmes wrote his father that the exchange business "is worth getting hold of" and predicted that "when this central office system becomes known there will be a good many bidding for it."[33] After his temporary excursion into this new technology, however, Holmes chose a project of more modest proportions and, as he aptly pointed out, of "more immediate pecuniary value." In place of an exchange business, he organized one of Bell's first "district systems," a combination messenger–private-line telephone service operated out of a single office. The district system was clearly an improvement over the unconnected operations of existing private-line services, but it would soon be supplanted by the fully interconnected central office.

Holmes's achievements on the banks of the Charles River were repeated a month later by Thomas Doolittle, of Bridgeport, Connecticut. Doolittle was, by his own admission, a "fair Morse operator," as well as "the instigator" and by far the largest owner of "a system of [telegraphic] communication exactly similar to a telephone exchange." An enthusiastic promoter of the telephone, as early as 1876, long before Bell's invention had become a commercial success, Doolittle had advocated substituting the telephone for Morse instruments on circuits maintained by the Bridgeport Social Telegraph Company. A year later, employing a modified switchboard and several lines belonging to the Atlantic and Pacific Telegraph Company, he connected together a number of locations, but he temporarily abandoned his experiments after the firm's facilities were closed down.[34] Although it was not until Charles Williams delivered a suitable switchboard (July 1878) that Bridgeport once again enjoyed exchange service, the publicity attending Doolittle's pioneering work in this field set the stage for the development of commercial exchange service elsewhere in Connecticut and throughout New England.[35]

To George Coy, of New Haven, Connecticut, has gone the distinction of establishing the first commercial exchange system. On 28 January 1878, Coy connected twenty-one subscribers using eight grounded iron wire circuits and a small, makeshift central office switchboard that he had constructed out of parts borrowed from his own district telegraph business. His exchange turned out to be an immediate success. After a mere three months of operation he was serving 42 residential and 185 commercial subscribers.[36] Before the books were closed on the first year of his New Haven operation, Coy celebrated the installation of his 405th station. By that time business subscribers had begun to urge their customers to "order by Bell Telephone," and perhaps in recognition of Coy's success, Hubbard had granted Coy permission to open additional exchanges in Hartford, Connecticut, and Springfield and Holyoke, Massachusetts. Technical achievement was matched by commercial success. Coy reported $11,302 in profits shortly after extending his activities into these new markets, and his accomplishment spawned widespread imitation within the ranks of Bell's licensees.[37] Thanks to Coy and his colleagues in Bridgeport and Boston, the market for Bell's telephone had been dramatically expanded. Technology had driven the business in a new, but not unforeseen, direction. The central exchange soon became the principal building block of integrated community telephone service.

Initially, however, there was still considerable uncertainty in the industry concerning the expense, design, and operation of the new central office systems. Holmes had briefly touched upon the basic economics of exchange service, the big unknown in the increasingly complex equation of telephone costs, in a review of the pricing and expenditure implications of his Boston experiments; but his estimates and forecasts had been speculative. He had predicted that a monthly rate of between five and six dollars would have to be charged for exchange services (in addition to the regular rental collected for the instrument).[38] Thus the total annual subscription charge would be between eighty and ninety-five dollars, depending on whether one qualified for Bell's residential or business instrument leasing rate. This was a considerable sum, and it would ensure that the market for switched services at first was confined almost exclusively to business customers.

One of the crucial new features of the exchange business was the high fixed investment represented by the central office plant, a condition that changed the economics of telephone service dramatically. From the licensees' perspective, the enlarged capital requirements and associated risks of exchange operations entitled them to a local monopoly franchise, or as Holmes described it, "an exclusive right to the use of [Bell] telephones . . . for all Central office purposes."[39] In the period of private-line development, facility costs frequently were recovered immediately when customers purchased their own lines. By contrast, the exchange represented a permanent capital investment only a portion of which was repaid through monthly service charges. The licensees building these central exchange systems were under considerable pressure to recover their fixed costs, and they did not welcome competition, which would almost certainly compel a reduction in rates and total revenues. There was also an obvious technological imperative favoring local monopoly. The crucial advantage of the switchboard was that it gave each customer access to the phones of all of the other customers in the same exchange area. With more than one switchboard, however, that advantage—or at least part of it—would be lost. Customers would have to subscribe to more than one service in order to enjoy complete coverage; facilities would be unnecessarily duplicated.

Watson began working feverishly on a design for a standard switchboard in early 1878, but his progress was slow, and he was repeatedly frustrated by the novelty and complexity of exchange technology.[40] In the meantime, Hubbard pressed the company's agents to furnish exchangelike district services, hoping to establish a loyal subscriber base that could easily be recast into a market for exchange services once Watson had completed his task.[41] By August, five hundred customers had signed up for Bell's district service in Boston, and hundreds more were patiently waiting for installation of the service in Lowell, Massachusetts; New Haven; Albany, New York; Nashville; St. Louis; Detroit; and Chicago.[42] After visiting several of these cities in the spring of 1878, Watson concluded that district operations would easily double the number of telephone subscribers for most Bell agencies; he later predicted that every major urban center in the nation would boast a district or exchange system within two years.[43] The agents were "working admirably and giving splendid service," the company reported in August, and by that time the district operations were already eclipsing Bell's private-line operations.[44]

These favorable prospects were suddenly qualified, however, when a powerful competitor, Western Union, entered the field. The telegraph company had acquired the rights to the telephonic devices of Elisha Gray and Professor Amos Dolbear, of Tufts College, and had commissioned Thomas Edison, a promising, thirty-year-old inventor from Menlo Park, New Jersey, to improve upon their elementary designs. In early 1878, Western Union began developing exchanges around the country.[45]

The beginnings of competition brought into sharp relief the deficiencies of the Bell Company's financial condition and the wide differences of opinion among the firm's executives over how to deal with this problem. Many of these differences came to the surface after Hubbard had, in typical fashion, unilaterally decided to push ahead quickly in the Chicago market by getting directly involved in the promotion of the exchange system. In late 1877, reports from the Windy City had indicated that the rental business was languishing, arousing Hubbard's suspicions about Anson Stager, his licensee, who coincidentally was also president of the Western Electric Company, the manufacturing arm for Western Union. This in itself was not an unusual arrangement, since Hubbard had routinely assigned franchises to candidates closely associated with or employed in the telegraph industry as a means for obtaining the talent and technical expertise necessary for building and operating a telephone system. However, questions concerning Stager's allegiance naturally arose after Western Electric agreed to produce telephones for the Bell Company's rival. After a visit by Watson (in the spring of 1878) failed to dispel doubts about Stager, M. H. Eldred, Hubbard's agent for Missouri, was hastily dispatched to take over and direct the Chicago exchange. Maintaining that the situation in Chicago merited extraordinary measures, Hubbard instructed Eldred to draw upon Sanders for funds when and if such support became necessary.[46]

Few initiatives in 1878 stirred up such a storm of controversy as did this one. What followed made the earlier deliberations leading to the formation of the New England Company appear as little more than a dress rehearsal for a larger struggle between Sanders and Hubbard over control of the firm. Predictably, Sanders argued for fiscal responsibility and criticized Hubbard for embarking on a policy that would overtask the Bell Company's finances. Hubbard, who still managed the enterprise as its trustee, maintained that a retreat from the Chicago market would undermine the confidence of the company's licensees; but he could not explain where Bell would get the funds to support the exchange business. A considerable amount of new capital was needed to sustain this operation, and as the bills mounted for the rescue in Chicago, the company was forced, in the beginning at least, to draw upon the personal accounts of Sanders to meet these expenses.

In March, Sanders wrote Hubbard about the "need of forming a company with capital not simply to supply . . . money for present wants." The company's weak financial condition had sparked rumors among the licensees that it would be unable to sustain the patents in a long battle with Western Union. Bell's New York licensee had already refused "to take any step until he [was] more confident in our financial ability," according to Sanders. Sanders went on to insist that "every man who looks at the matter but yourself sees the absolute necessity of an organization with capital, and when you weigh it carefully I am convinced that you will come to the same conclusion." The firm's financial weaknesses were compounded by the fact that collections were running behind schedule. For the month of February, Sanders reported a business of only $2,804 outside of New England. The company owed Williams $6,000. "Borrowing money will not help us out . . . as the pay day must come sooner or later," he concluded. "The moral effect of the knowledge that we are strong and not borrowers will . . . strengthen our agents, causing them to collect and remit money. The actual facts of the case are that, as sensible businessmen our agents at this stage are cautious about involving themselves with us."[47]

But the idea of bringing new investors into the business again aroused Hubbard's opposition. Not only did this move entail an implicit negative judgment of his performance as the firm's chief executive (by now a matter of some controversy) but it would almost certainly require the original patentees to surrender another portion of their authority over the business. The Bell Company's trustee was unhappy on both scores; however, he also recognized the seriousness of his company's financial condition. If the admission of new partners was necessary, Hubbard was determined to avoid inviting into the reorganized venture the same group of Sanders's friends that had so quickly and effectively clipped his managerial wings in the New England Company. Instead, he enlisted the support of his company's Philadelphia representatives in determining whether investors in that city were willing to buy $300,000 worth of Bell stock at a 50 percent discount. His proposition met with indifference. Philadelphia investors did not feel secure investing in an enterprise "so far away." The city's financial community had just been "badly 'bitten' by the heavy decline in railroad stock."[48] Subsequent attempts to obtain support from a syndicate of local agents were also unsuccessful.

In the meantime, Sanders had opened his own round of negotiations with Bradley, hoping to win his support for a recapitalized, consolidated national enterprise. Throughout the early spring of 1878 a variety of formulations were discussed, but the parties could not agree as to how much authority the monied interests and the original patentees should have. Hubbard's recalcitrance on matters affecting his future position within the reorganized venture proved to be the greatest stumbling block. With each new proposal, he insisted on modifications limiting the influence that Bradley and his friends would have over the new business.[49]

His tenacious opposition eventually paid off. Hubbard's frustrated opponents finally agreed to form a new, separate company out of the original Bell Company and to concentrate its commercial activities beyond the New England market. This arrangement generally satisfied Hubbard, who thought that he would exercise more authority over the affairs of the overall business of two smaller companies than he would have had they been organized as a consolidated, single firm. Moreover, Bradley and his colleagues could be counted on to supply some much-needed capital when they purchased a position in the reorganized company, which was formally incorporated as the Bell Telephone Company in June 1878.[50]

Hubbard's victory was not, however, unqualified. Although after reorganization the patentees wound up holding nearly three-quarters of the new Bell Company's $450,000 pool of stock, the management structure of the firm was established in ways that greatly reduced Hubbard's authority.[51] The investors won equal representation on the company's board of directors; and the board's Executive Committee, which now comprised Thomas Sanders, Gardiner Hubbard, and George Bradley, was not likely to vote for the kinds of policies that Hubbard had been following. Even on matters of a routine nature, Hubbard's authority as president had been circumscribed. Responsibility for managing the business on a daily basis and for controlling licensee contracts and relations was vested with the company's new general manager, Theodore N. Vail, a little-known but highly regarded former superintendent of the nation's Railway Mail Service.[52]

The appointment of a new general manager from outside the ranks of the investors and patentees heralded a significant departure in the administration of the firm. Until this time, the Bell enterprise had been directed by entrepreneurial leaders whose outside commercial interests and political obligations had regularly diverted their attention from telephone affairs. Now the company's first salaried, full-time general manager would supervise the daily affairs of the business, conduct regularly scheduled tours of the field, ensure the diffusion of technical information among the licensees, and handle corporate correspondence. Along with Oscar Madden, who was recruited about the same time from the Domestic Sewing Machine Company and placed in charge of agency affairs, Vail would launch the Bell enterprise on the road to professional management.

Although Hubbard continued to enlist local candidates for agency assignments and, with his colleagues on the Executive Committee, to formulate policy, his principal responsibilities after Vail arrived were limited to legal affairs. No longer was it possible for him alone to set policy for the firm. The organization was now more than the product of any single individual. The business was now an incorporated body, and its Executive Committee symbolized the new distribution of power and the more formal ways in which policy would be determined and implemented in the future. Vail and Madden were in many ways the agents of this new structure of authority. Together they would impose order on a business that had grown in a haphazard, ad hoc fashion. The administration of the enterprise would soon exhibit the makings of an orderly decision-making mechanism, organized around a new collective leadership, based upon notions of consensus, cooperation, and a systematic division of administrative responsibility. Its board would furnish a vehicle for managing routine transfers of executive power, giving the firm a permanence and stability that it had not enjoyed under its entrepreneurial leadership.

The new organization also counted among its assets greater administrative and financial resources than in the past, but these would be sorely taxed in the years ahead. Even the new Bell Company amounted to little more than a dwarf when compared with Western Union. The task of developing an organization well suited to the new demands of the exchange business had only begun. Change was in store for the company's relationships with its licensees and agents, the manner in which expansion was funded, and, above all, the company's strategy for dealing with its new, powerful rival in the telephone business.

CHAPTER 3 ❦

The Uncertain Years: Competition and Administrative Consolidation

---

THEODORE N. VAIL—NO OTHER OFFICER in the history of the Bell System would come to inspire as much admiration among employees and managers alike. Vail was the most unusual of the corporate leaders who over the years left their mark on the Bell enterprise; certainly he was the most revered. Industrial statesman. Articulate spokesman and defender of his firm's interests before the public. Advocate of regulation. Entrepreneur. Monopolist. Vail more than any other man would shape and personify the modern Bell System.

Vail was only thirty-three when he accepted Hubbard's invitation to become the Bell Company's first full-time general manager (July 1878). On the surface at least, his appointment was a bit surprising. He had little in common with his Bostonian employers. Born in Minerva, Ohio, in 1845, the son of a Quaker farmer and ironworker, Vail had spent much of his early life enrolled in what one of his biographers described as "the school of versatile experience."[1] After moving to New Jersey with his family at seventeen, he had learned the art of telegraphy while working as a clerk in a small-town drugstore. Two years later, as an operator in Western Union's New York City office, he had had his first encounter with the world of big business, a world to which he would later make a significant contribution.

His career at Western Union was temporarily interrupted when his parents left the secure environs of New Jersey for the frontier of Waterloo, Iowa, in 1866. There, the hearty but uncertain life of the farmer became his mainstay for the next two years. But in 1868 he returned to the telegraph business, this time as a night operator for the Union Pacific Railroad at Pinebluff, Wyoming.

Vail's career took its first sharp turn when he left the railroad to work for the United States Mail Service. He became an ardent advocate of administrative efficiency and devised numerous plans to improve the performance of the service's operations. His initiative and drive won him a transfer to the Washington office of the Railway Mail Service, where, after becoming assistant general superintendent, he inaugurated a much-acclaimed Fast Mail Service between New York and Chicago over the tracks of the New York Central and Hudson River railroad systems. In recognition of these achievements, Vail was appointed the service's general superintendent in 1876.

While serving as superintendent of the Railway Mail Service, Vail came to the attention of Gardiner Hubbard. Hubbard, a congressional advisor with an active interest in postal affairs, saw in the young, energetic Vail just the type of administrator his small but fast-growing telephone company needed. In 1878 he persuaded Vail to relinquish his secure position in the mail service for the uncertainties of managing a firm dangerously close to financial insolvency. But for Vail the business of telephony offered opportunities for growth and a challenge that few other jobs could match. For Hubbard, the hiring of Vail as the Bell Company's first general manager would be his last significant contribution to the business as its chief executive—but it was a decision that would influence the firm and the industry well into the next century.

When Vail joined the Bell Company, the industry was undergoing a sweeping transition. The tinkerings of a small circle of ingenious New England licensees had led to the establishment of the first central office systems, an event that suddenly shifted the locus of technical development out of the company's hands and prompted a hectic effort on the part of Watson and Charles Williams to stan-

Theodore Newton Vail, the first general manager at the Bell Telephone Company, appointed in July 1878. Vail continued to hold the position of general manager in the reorganized National Bell Telephone Company and its successor, the American Bell Telephone Company. He resigned this post in June 1885 and assumed the presidency of American Bell's long-distance subsidiary, AT&T, on 14 August 1885. In the fall of 1887 he left AT&T to become president of the Metropolitan Telephone and Telegraph Company, Bell's New York licensee. In 1889 Vail resigned from an active role in the business, though he maintained a position on AT&T's board of directors until 7 May 1892. A decade later he would return to AT&T's board, this time as a representative of the major banking interests associated with J. P. Morgan and Company. Vail, the company's first full-time, professional manager, is credited with systematizing Bell Telephone's licensing arrangements and with introducing methods of administration that greatly improved overall efficiency during the formative years of the firm's development.

William H. Forbes, National Bell Telephone's first president. Forbes, the son of a New England railroad magnate, led an investor revolt that ultimately removed Hubbard and Sanders from important positions of authority within the Bell enterprise. He became American Bell Telephone's first president when the business was reorganized and recapitalized in 1880. Under his direction, the company reached a pathbreaking patent settlement with Western Union in 1879, one that restored his company's monopoly over telephone technology. Although he resigned the presidency in 1886 to pursue his other commercial interests, Forbes retained an influential position on the board's Executive Committee into the late 1890s.

dardize (and presumably patent) an efficient switchboard for use in the field. An even more disturbing challenge to the Bell Company's technological leadership came from another quarter. Not at all pleased with the vigorous manner in which the telephone business was growing, Western Union now saw in Hubbard's enterprise a distinct threat to its control over the lucrative message business. William Orton, the telegraph giant's president, had harbored a grudge against the management of the telephone enterprise ever since Hubbard, in his capacity as advisor and principal investigator of the telegraph industry, had urged the nationalization of Western Union's intercity operations. Orton now decided to enter the field. He hired a team of leading scientists—including Edison—to develop patents and devices that would enable his company to compete in the new field, and in late 1877 Western Union launched through its subsidiary (the Gold and Stock Company) the American Speaking Telephone Company, a formidable entrant into the telephone sweepstakes.

Over the next two years the two rivals engaged in what was later described as a "race to occupy the field," a contest premised on the notion that the first to establish a central exchange at any given location enjoyed crucial advantages over those who arrived later. Price competition was not a tactic chosen by either contestant, however. Indeed, Western Union, at least, seems to have considered the struggle with the Bell Company only a temporary affair, one that would eventually lead to an amicable settlement of differences and perhaps even some form of consolidation. Under these circumstances, a price war made no sense. Raising subscriber charges would be difficult enough once monopoly was restored. Meanwhile, the race continued, and legal proceedings over the disputed patents began. The fight would take place within what Western Union officials hoped were well-established boundaries.

How well did the Bell Company stack up against its rival? Surprisingly well. Bell's claim on the basic telephone technology was relatively secure—an opinion shared from the outset by counsel for both parties. Two months after Hubbard filed an infringement suit in the Massachusetts Circuit Court against Gold and Stock's local agent in Boston, Western Union's chief attorney warned his colleagues that "on the broad ground of the methods of telephone transmission," he expected Bell's patent to prevail. Elisha Gray, the scientist responsible for building Western Union's first competing telephone, had not "reduced [his] invention to practice" before Alexander Graham Bell had been awarded the telephone patents.[2] Edison's later success at devising an efficient carbon transmitter confused the issue somewhat, but in preliminary hearings the court seemed inclined to view such innovations as derivative rather than original. In any case, by the end of 1878, Francis Blake had designed what Hubbard and Sanders believed was an even better "variable carbon transmitter" for the Bell Company. Western Union's patent claims were uncertain at best.

In the field, however, the Bell Company's position was considerably weaker. There the struggle would be won by the swift, in this instance the enterprise whose financial resources could sustain a broad and rapid drive to occupy the best markets. Because of its limited capital resources, the Bell Company was at a distinct disadvantage. Western Union counted assets approaching $40 million; this fact alone gave the firm and its subsidiaries—Gold and Stock and American Speaking Telephone—the financial muscle and credit that they needed to fund expansion. Hubbard had made the most of his less enviable position by enlisting an expanding contingent of entrepreneurs as local representatives and agents. But it was questionable whether this loosely organized association would make up for his company's limited resources. Certainly a more coordinated and better-funded effort in the field was needed if the Bell interests were to be considered serious contenders.

The task of strengthening the Bell Company's position fell to Theodore Vail, who was confronted with serious, immediate problems from the day he assumed his duties as general manager. At the first indication of Western Union's aggressive intentions in New York City, the company's two local agents had abandoned their commitment to erect an urban exchange, citing unanticipated costs and Hubbard's failure to defend them against competition as reasons for their decision.[3] In Buffalo, New York, then a major gateway to the Midwest, the situation was no less difficult. There an anxious Edward J. Hall, fearing a determined thrust by Western Union into his market, petitioned Vail for some financial support. Just the rumor that the giant telegraph company was on the move was enough to unsettle Bell licensees. Much of the company's time and energy had to be spent conveying a spirit of confidence and commitment in an attempt to bolster the morale of its local agents. But assurances concerning the validity of Bell's patents could do nothing to obscure the obvious financial shortcomings of the enterprise. Agents had to wonder whether Hubbard and Vail would successfully withstand the strain of a prolonged contest in the field and in the courts.

Vail could do little to improve his company's financial standing, but he could help generate confidence among its licensees by removing many of the inconsistencies embodied in the policies formulated under Hubbard's management. To licensees in the hinterland, isolated from the home office, abrupt and unexplained shifts in policy were interpreted as a manifestation of official indecision in the face of competition. Inconsistency in policy also undermined the overall sense of community holding the Bell Company and its local agents together as a unified enterprise. As general manager, Vail tried to iron out some of the differences arising out of the licensing arrangements that Hubbard had personally negotiated. He also tried to work out a coherent policy for dealing with licensees' requests for financial support.

Given Bell's financial condition, the company had no business contemplating financial support for the local operations. No promises to that effect had been made when the licenses were first granted. The company needed all of its resources for funding the production and distribution of its instruments and for defending its patents. But competition drove the Bell enterprise in this new direction. It was not easy for either Hubbard or Vail to remain idle when the agents were struggling to best the powerful and determined Western Union Company, especially after the Bell Company's notable losses in New York. In June, Hubbard announced, without first seeking his board's advice or consent, his intention to extend financial aid to the licensees in return for an interest in their exchange operations.[4] His decision surprised and angered his colleagues. The policy seemed too risky, too expensive, too arbitrary.

Once such investments had been made, however, it was difficult to abandon them. As Vail explained to Bradley (in reference to a case similar to the one they faced in Chicago):

> No one can more fully coincide with you than myself in regard to this Company building itself up in guarantees that may possibly

show more largely than estimated upon the resources . . . yet after
it has become involved I think it just as disastrous to withdraw from
the fight and let the opposition take possession of the field. The
mistake this Company made . . . in regard to the Buffalo business
was in starting the canvassors before the capital was obtained. . . . if we still go further and remit to them all the first years
rentals it may be by far better than to let the opposition have full
possession of the City of Buffalo and surrounding country. This if
we intend to make ourselves a permanent organization and live
beyond the time our patents may run.

In the face of such uncertainty, the company could not afford to sit
on the sidelines, particularly in its most important urban markets.
More than moral support was required if the licensees were to be kept
from defecting to Western Union.[5]

Sanders, who was the company's chief financial officer,
fought Hubbard's decision to invest in the licensees. He thought that
the Bell Company, being a financially undernourished enterprise,
lacked the capital needed to underwrite this new policy. Besides,
direct ownership of local operations would almost certainly subject
the company to large legal penalties if it lost the impending patent
suits. But even Sanders was flexible when it came to extending relief
in the form of concessions on instrument payments. In a few cases
the company even eliminated royalties altogether for the first two
hundred or so telephones that designated licensees installed.[6] In this
way the Bell Company improved the working capital and cash
position of the local businesses. But it was later determined that
despite Hubbard's earlier suggestions, the company would not itself
go directly into the business of establishing exchanges; its "settled
policy" was that "all District Companies [would] be run by local
agencies."[7] Hubbard reluctantly agreed. "Our plans," he told the
licensees, "are to organize these companies in all the large cities and
throw upon the residents the responsibility and profits of the business."[8] In Chicago and New York, Hubbard found it difficult to
escape the obligations that he had already incurred.[9] For the most
part, however, the firm would at first follow the cautious course that
Sanders wanted to steer.

Vail, for his part, sought to ensure that the license contracts
evenly matched agency resources with the expanding needs of the

central office business. Many of the agreements signed before the development of the central office system had been specifically fashioned with the market for private-line operations in mind. It was not uncommon, for example, to find agents holding franchise rights to comparatively large territories, in some instances comprising more than one state. Given the substantially higher costs of the exchange business, however, the agents could not fund a rapid expansion of Bell services into so large an area, and thus room would be left for Western Union to move in and occupy the market.

Vail set out to renegotiate such agreements. Frequently, agencies were, of course, reluctant to yield any part of their territory. Charles Haskins, who held franchise rights covering the state of Wisconsin and elsewhere, was a case in point. Hubbard confided to Vail that the company would "find it difficult to get [the original franchise] away from him" and warned that in the future "any man that has so large a territory will prove hereafter an incubus" on their plans for introducing exchange services on a large scale.[10] But even Haskins finally gave way. He and a number of the other licensees in effect traded a large territorial franchise oriented to private-line development for a more compact franchise oriented to the city-centered exchange business. As Hubbard explained to those disappointed with the new policy, "The exchange business, which is the principal work your district company will probably perform, was not in operation or thought of when our contract was made with your Company. It is rapidly becoming a most important part of our business and the most profitable to our customers."[11]

Where the agents were unwilling to budge, Vail and Hubbard backed off, leaving the issue of franchise boundaries to be solved at a later time. But in most instances Vail achieved his goal of tailoring the franchises to fit more sensibly the new technology and economy of the central exchange.

At the same time that Vail was reshaping agency assignments, he was seeking to establish a higher degree of uniformity in the firm's royalty and commission policies. Hubbard's highly personalized method of administration had left behind a legacy of extraordinary diversity. Licensees had asked for and in some cases received a wide range of rental concessions from Hubbard. But Vail was generally opposed to this approach, insisting that "to do business at a less figure [than five dollars net on each instrument] except for special reasons was as bad as not doing . . . business at all." Vail doubted that deep royalty reductions by the Bell Company or local price cuts in service constituted the best way to meet competition over the long run; in several instances, Vail believed, "exchange companies had been charging less than enough to enable them to give first class service."[12] Poor service would eventually undermine the company's position, leaving it vulnerable to competition.

As a means for dealing with these irregularities in policy and practice, Vail drafted a standard license contract and persuaded the firm's officers to adopt it.[13] The terms under which the licensees obtained instruments were standardized, as were the prices that licensees were to charge customers for the privilege of leasing Bell telephones. The Bell Company was to have first rights to acquire any agency telephone properties if agents chose to leave the business. This clause protected the company against Western Union's efforts to expand its own business by purchasing competing agencies. The standard contract represented an important accomplishment for Vail, who at this time had been with Bell for only six months. In this and other ways, he sought to establish a new uniformity throughout the enterprise so that it would be better able to cope with competition.

Given the internal and external difficulties that the company was facing, it is surprising that it did as well as it did during Vail's first years in office. Actually, 1878 was a year of rather impressive gains for the Bell interests. The report submitted to the board of directors in February 1879 described the swift spread of exchange and district services along the East Coast as far south as Richmond, Virginia. Beyond the Alleghenies, in the more sparsely settled regions of the Midwest, development had been much slower. But in the entire nation, Bell agents had installed an astonishing twelve thousand telephones over the past twelve months (each providing the company with a net rental fee of about five dollars). With the recent opening of exchanges in Utica, Syracuse, Cleveland, Nashville, and Indianapolis, demand for instruments was expected to approach from fifteen hundred to two thousand units a month during the spring of 1879. Excluding the earnings of the New England Company, Bell recorded a profit of twenty five thousand dollars on revenues of only thirty four thousand dollars.[14]

However respectable the company's income statement was, it was more than offset by the unfavorable situation in its accompanying balance sheet. The firm's capital reserves had been nearly exhausted by investments in Chicago. Thirty thousand dollars, about 40 percent of the paid-in equity, had been spent to fund the construction and expansion of exchange facilities in that city alone; the company owed another ten thousand dollars for planned extensions. These expenditures were deemed necessary in order to preserve the company's expanding position in the market and its reputation among affiliates. While licensees had made some notable progress in occupying the field, thus far their financial performance had been relatively poor. The start-up costs for central office development were higher than anticipated for most of the licensees. For the near term, at least, the prospects of achieving a better return on their telephone investments appeared slim.[15]

The adoption of new and more efficient methods of administration and improved coordination would help, of course, and with this in mind, Bell officials considered hiring a pool of permanent, salaried traveling representatives to convey information concerning the latest developments in telephone technology and corporate policy to licensees in the field. Two agents had already been employed on a trial basis in western New York and in the South, and their efforts in these areas, as well as their success in recruiting new local investors, appeared to justify the call for hiring more.[16] But short of capital, management determined that an improvement in the firm's financial condition was required before such developments could take place.

In the fall of 1878, the company's investors had already begun to consider such a change. Convinced that the separation of the New England and Bell companies made little sense in light of the growing demands of competition and the increasing financial requirements of the business, some argued for a reorganization and recapitalization of their interests based on a combination of the enterprises. It was originally suggested that the two companies might maintain their discrete financial functions under a single treasurer, but this half-measure was soon rejected. Bradley and his colleagues were reluctant to support such a scheme unless Hubbard agreed to relinquish overall responsibility for the consolidated company's financial affairs.[17] Not surprisingly, this precondition was greeted with little enthusiasm by Bell's ambitious trustee, but Hubbard's opposition was tempered by his recognition that the firm obviously needed an infusion of new capital.

Sanders pressed hard for a complete consolidation. He pointed out that the heavy investments of the New England Company in Boston district operations left it particularly vulnerable to an assault by the telegraph interests. Consolidation would spread the New England firm's risks and give its investors a stake in markets around the nation. These markets, Sanders thought, were going to grow more rapidly than markets in the Northeast. He also thought that the Bell Company would be unable to move effectively into the intercity market unless there was a complete unification of assets.

> As the business develops and the various district systems established in all parts of the country become connected by wires and the transmission of messages for hire is actively engaged in which it soon will be, we shall then realize how great an advantage it will be to have one common interest and to be confined to no sections. This latter point I think is of great importance. Our contracts with Agents and District Companies as well as those of the Bell Telephone Company invariably provide for the office facilities in the receipt and delivery of messages for hire. Such facilities are one of the largest items of expense to the W.U. We have the sympathy and support of the public and local capital will no doubt furnish these connecting lines on such terms as will give us the control of a system which will be a most powerful opposition to the WU system even if the telephone patents are not sustained.[18]

Sanders's thoughts were by no means original, nor did they anticipate what the nation's next generation of telephone users would come to know as the Bell Company's long-distance system. His was a vision of a Bell-built telegraph network that in combination with the company's phones would compete with Western Union for the lucrative business message trade.

While this proposal was being discussed, Hubbard's position within the enterprise was becoming increasingly controversial. In December 1878 he failed to present New England investors with an accurate account of the Bell Company's financial condition.[19] An audit ordered by the board of directors subsequently disclosed that the firm's debts had been mounting to a dangerous level. Although the board quickly passed an emergency resolution authorizing the sale of six-tenths of the company's Chicago holdings in order to help preserve the firm's cash position, the incident surely undermined confidence in Hubbard's administration of the enterprise.[20]

Hubbard himself recognized that the job of managing the company's affairs had become too complex for one with so many other obligations. He candidly confessed to Madden how difficult it was to keep track of the business while tending to his interests in Washington, D.C.[21] Many of his colleagues also had outside commercial investments, and as a result, policy was often formulated in an unsystematic fashion. Later, William Forbes, who joined Bell's board of directors in late December 1878, would point out that "it was attempted to do the business through an executive committee of three, one living in Boston with the New England Company upon his hands, one in New York, and the third the President in Washington also with many other affairs to attend to. . . . Mr. Hubbard tried to manage the Company's business, as President, without having anything to do with its finances or keeping the run of them . . . which, is to men conversant with such matters, rather startling."[22]

Forbes was to be one of the architects of the consolidated enterprise, and he brought to the board a keen appreciation of the problems of organizing and running a large, complex enterprise. He had been associated with a number of railroad companies and was familiar with their systems of administrative and financial control.[23] He helped introduce a similar form of bureaucratic control into the telephone enterprise. The key feature in the plan that he submitted to the board involved the formation of a new, three-member executive committee. Vested with authority over the "executive functions" of finance and policy development, the committee was to be a full-time "working group" made up of experienced businessmen and elected by the full board. This would centralize responsibility and provide a new form of collective leadership. The consolidated company's board would be increased to eleven members and its authority over the appointment of all officers (except the treasurer) strengthened.

Forbes intended to exclude the patentees from the new Executive Committee and to install an entirely new, professionally qualified administration to restore investor confidence. He explained to his colleagues:

> To make Mr. Hubbard President and put him on the committee would be to practically continue the old conditions of affairs, and this would not be satisfactory to the stockholders.
>
> With feeling entirely friendly to Mr. Hubbard [investors] cannot help feeling that he lacks that business training which the Company's affairs imperatively demand, and that nothing so proves the undesirability of placing the practical part of the management in his hands as the fact that he now believes that the business has been well managed, and that it is successful, and its success due to his management.[24]

Stung by the recent financial revelations, Hubbard himself seemed inclined to concede the "propriety" of conferring "a controlling interest in the business [on the investors] until its success was fully established."[25] But he demanded more than an honorific role in the new organization. Together he and Alexander Graham Bell would still hold about one-third of the company's outstanding shares, even after consolidation, and for that reason alone he was opposed to accepting merely a token presidency. In protesting to Forbes, Hubbard was joined by Thomas Sanders, who by now had his own misgivings regarding his fate in the reorganized company.[26]

Appropriately enough, it was upon the shoulders of Alexander Graham Bell that this stalemate finally came to rest. Outwardly indifferent to the commercial affairs of the business, Bell had maintained an enduring personal interest in the telephone as a triumph of science. He was viewed as even-handed and just and enjoyed both the respect and the confidence of company investors, as well as close personal ties with Sanders and Hubbard. His opinions carried weight among all those who saw in him the founding genius of their enterprise.

However much he appreciated the merits of placing the company in the hands of more experienced businessmen, Bell declined to support a plan that would completely remove his father-in-law from a position of managerial responsibility. He urged the prospective members of the new Executive Committee to seek the "assistance and advice of those who are acquainted with the past history of the business," but he stopped short of recommending that the investors welcome both Sanders and Hubbard as equal voting members of that body. They deserved, he believed, enough of "a consultative voice" in company matters to relieve him of any anxieties he harbored about turning those affairs over to "men who have only a pecuniary interest in the telephone." So long as his "personal friends" remained involved in the management of the firm, Bell said, he would be confident that the corporation's policies would not in any manner "compromise [his] honor or reputation."[27]

Both sides accepted this compromise, and in mid-March 1879 they turned their attention to the basic issues of implementing the combination. Working committees staffed by both New England and Bell shareholders labored over the details of consolidation. In anticipation of the reorganization, an agreement of association had been submitted to the General Court of Massachusetts in mid-February; the agreement notified state officials of the impending formation of a new enterprise, to be incorporated as the National Bell Telephone Company.[28] Plans were made to issue $850,000 in capital stock, slightly less than originally contemplated, and a new headquarters was established in Boston. The company's proposed charter was drafted in the broadest terms; it allowed management wide discretion in developing the market for telephone services. The business was variously described as the manufacture and leasing of telephones "and their apparatus" and the construction, maintenance, and operation of "lines for the transmission of messages by electricity or otherwise."[29] On 13 March the firm's fourteen shareholders toasted the news that the governor's office had approved their petition for incorporation.

At a meeting on 10 March the shareholders had voted formally to expand the company's Executive Committee to five members, and on the following day the corporation's eleven directors had appointed William Forbes, Gardiner Hubbard, and Thomas Sanders, along with R. S. Fay and inventor Francis Blake, to that body.[30] The directors elected Forbes president; installed George Bradley as treasurer; and confirmed Theodore Vail as general manager and Charles Hubbard as clerk. Of the 8,500 shares of new stock then issued, 2,000 were exchanged on a one-to-one basis for equity held by New England Company investors; another 4,500 certificates were earmarked for the original Bell Company stockholders. An additional 750 shares were sold on the open market to raise badly needed working capital to cover the debts that the new firm assumed from its corporate predecessors. That left 1,250 shares in reserve for future financing.[31]

With the formation of the National Bell Telephone Company, the Bell interests opened a new chapter in the development of their enterprise. But the new administration would have little time to relax and celebrate this occasion. Already there were signs that its giant adversary was interested in negotiating an end to competition and the impending patent contest in the courts. Meanwhile the industry's rapidly changing economic and technological environment was creating new opportunities and new challenges for management. During the first nine months of his presidency, Forbes would be forced to make a number of crucial decisions that would have a lasting impact on the telephone business.

CHAPTER 4 🐦

Accommodation with Western Union

WITH THE ORGANIZATION OF THE National Bell Company, the Bell interests were in better shape for the next round of competition with Western Union. Gardiner Hubbard's energetic but relatively haphazard direction of the enterprise had given way to a more controlled brand of leadership. Vail's efforts to systematize the business would now, it seemed, be fully supported by the rest of the company's officers and by its stockholders. National Bell at last had a strong enough foundation—with the New England business and that in the rest of the country under one corporate control—to promote with vigor the expansion of the exchanges. Much needed capital had been acquired.

But then, just as the next round of competition seemed to be under way, National Bell and Western Union reached a settlement. Neither the forces of technology nor the pressures of the market appear to explain what transpired. Yet the agreement was the single most important factor influencing the industry and National Bell's position in it for more than a decade. Without it, the National Bell Company would not have enjoyed the field of telephony to itself for the next fifteen years. To this day, however, it is unclear just exactly why the nation's biggest company surrendered a promising market to its small competitor.

Several explanations can be advanced. For one thing, the Bell interests enjoyed what had come to be viewed as a very strong patent position in telephony; this certainly drew Western Union in the direction of accommodation. Then, too, there were outside forces threatening William H. Vanderbilt's control of the telegraph colossus, making him more amenable to a quick solution of the telephone problem. Western Union's size and organization also may have contributed to the lack of entrepreneurial vision that its management seems to have displayed at this juncture. Perhaps the longstanding desire to monopolize telegraphy simply overshadowed management's interest in the development of a related but different market. At any rate, the reasons for the settlement were complex, and the process of accommodation needs to be examined with some care.

In the latter part of the nineteenth century Western Union was one of the nation's few modern corporations outside of the railroad industry. Its size and professional management placed it at the head of the field in electrical communications, which it had dominated since its founding. It had led America into a communications revolution, and its sprawling wire network was the crucial means of linking the nation's leading centers of commerce and industry. By late-nineteenth-century standards, Western Union's management had an enlightened appreciation for technical innovation; it routinely monitored and acquired inventions in the advancing science of electrical transmission and reception.[1] Its business was, by and large, the transmission of business messages and news; its history, one of vigorous corporate acquisition aimed at creating and maintaining a single national system of communication.

But ample as Western Union's resources were, they provided no sure defense against Jay Gould, by reputation America's most ruthless financier. Gould had distinguished himself by mounting a series of highly visible assaults on corporations, largely those in transportation. Skillfully manipulating both courts and legislatures, he had exhibited few if any compunctions about the means he used to achieve his immediate ends. Now the nation's leading "robber baron" was apparently determined to establish a rival telegraph system using the newly built Bell exchanges for local connections.[2] In the past, when threatened by competition from a regional carrier, telegraph company management had either negotiated an end to such rivalry or launched a price war aimed at weakening the opposition prior to buying them out. These tactics, however, probably would not have sufficed to defeat Gould, a worthy and dangerous opponent. His attack—his second such foray in a short time—no doubt sent shivers of anxiety through the ranks of Western Union officials and must have renewed an interest in an accommodation that would prevent a Bell-Gould alliance.

Western Union officials had for some time been looking for a basis on which to achieve a consolidation with the Bell interests. Not long after Western Union's American Speaking Telephone Company had begun soliciting subscribers, telegraph officials had convened a series of secret meetings with Charles Cheever, Bell's New York licensee, to explore the prospects for an agreement.[3] William Orton, who presided over the telegraph company at this time, offered to capitalize a new, combined enterprise at an astonishing $3 million and staff it with experienced Western Union officials. The Bell interests would be given a generous $1.00 payment on each telephone rented by the joint venture, a 7 percent return on their portion of the investment, and a minority position on the board.[4] This was a serious offer, one not lightly dismissed by the officers of a small, financially strapped enterprise.

But neither Hubbard nor Sanders had much faith in Western Union's commitment to telephony, and both had strong reservations about relinquishing their control over the valuable Bell patents. Even though the proposed partnership would have eliminated their vexing financial difficulties, Sanders concluded that a complete sellout was preferable to any cooperative arrangement in which Western Union assumed a dominant role. Perhaps unrealistically, he hoped to avoid a major confrontation, convinced that the Bell Company could eventually find a profitable and secure niche in the market without encroaching on the primary interests of its mammoth adversary. Hubbard, on the other hand, seemed a little more receptive to Orton's idea, but he (predictably) refused to sanction any accommodation that would in effect displace him as president.[5] He demanded a pledge from Western Union that the development of the telephone business would not in any way be sacrificed in the interest of its telegraph operations. Specifically, Hubbard sought a guarantee from Orton that the consolidated enterprise would install fifteen thousand telephones during its first year and twice that number the next.

Terms such as these were not welcomed by Western Union, and both sides soon decided that they had little common ground upon which to base an agreement. After two months of discussion, they continued to disagree over the administration and control of the new enterprise. The meetings ended on a sour note, with Western Union officials threatening to install competing phones in every post office and telegraph dispatch center across the nation; Hubbard ruminating about a devastating patent suit. An all-out battle was averted only after Orton ordered his competing agencies to price their services in line with the local schedules posted by their Bell counterparts.[6] In this way he left the door ajar for further negotiations (and also protected his own agencies' profits).

Orton's decision did not mean, however, that the Bell interests would be shielded from competitive pressures as Western Union advanced into telephony. At times throughout 1879, rate cutting broke out—a development that forced National Bell to undertake what Vail later referred to as "a little judicious dead heading."[7] By this he meant that the telephone company sometimes reduced the price it charged licensees for transmitters, at other times furnished phones rent-free for brief periods, and in some cases even guaranteed the costs of connecting a certain number of subscribers to exchanges as part of its effort to help its affiliates meet Western Union's competition.[8] Normally, however, an undisciplined price war along a broad front was eschewed as both parties raced to "occupy the field." The first to establish a working exchange, most believed, enjoyed a distinct advantage over the competition, notably a liberally construed municipal or town franchise and with it, a secure position in the market. This style of rivalry was not as debilitating to the Bell interests as price competition; nevertheless, it severely strained the company's resources.

The climate for a reconciliation improved markedly after both Western Union and the Bell Company came under new management in 1879. Novrin Green took on the presidency of the giant telegraph monopoly following the death of William Orton, and William Forbes assumed the helm at the Bell enterprise after the organization of the National Bell Company. From the outset, both men seemed determined to negotiate an end to their rivalry. Green looked upon the reopening of negotiations as a means of forestalling the activities of Jay Gould; Forbes viewed a settlement as a much-needed respite from increasing financial pressures. Thus, National Bell officials were receptive to Green's appointment of Dr. Samuel White, the American Speaking Telephone Company's largest private investor, as his personal envoy at the revived discussions in April 1879.[9]

In his first meeting with National Bell representatives, White proposed "a plan of union" and invited Forbes to "perfect more carefully" his own position regarding such a consolidation.[10] Although National Bell's new president at first believed that it was possible to devise a plan that would protect the interests of the telegraph corporation without unduly compromising "the legitimate fruits" of National Bell's own labors, barely a week after opening these discussions he was forced to conclude that Western Union did "not care to consider any plan which would not place the control, by majority interest, of the Telephone business in their hands."[11] Over the next several weeks, company representatives nevertheless continued to discuss formulas for dealing with the management structure of a consolidated business. The last such proposal involved the establishment of a $5 million company. Three-fifths of its equity was to be divided evenly between National Bell and the telegraph interests, and a board of arbitrators would distribute the remaining $2 million in stock on the basis of the final verdict in Bell's patent case.[12] Forbes seemed encouraged by this turn of events.

For the telegraph interests, White's proposal of mid-May had one overriding feature; it would preserve the status quo in Western Union's message markets. Writing to Forbes on 3 May, White signaled his willingness to abandon the exchange business completely to the new consolidated venture as long as Bell agreed to restrictions prohibiting the use of its telephone facilities for the transmission of news dispatches, stock and gold market quotations, and business messages for hire. Later White clarified his company's position: "It must be understood and agreed as a condition precedent to the settlement proposed . . . that the new Company shall contract not to compete with Western Union Telegraph Co. in its telegraph business, nor to connect with or give business to any rival telegraph company."[13]

By early June it appeared that an agreement on a basic framework for consolidation had been fashioned. On 6 June, Forbes sent White a lengthy memo outlining his understanding of the general plan and suggesting that both parties review each other's existing contracts and liabilities before final arrangements were completed. White proposed a conference to iron out the minor differences that remained and to clarify a number of vague points in the agreement. Noting that the laws of the state of New York had "much more desirable features" than did the laws of any other state, he recommended that the new company incorporate and establish headquarters there. Because the state's existing statutes granting right-of-way privileges dealt with telegraphy, not telephony, the word *telegraph* would have to appear in the company's title. The Consolidated Telephone & Telegraph Companies of North America was an appropriate name for the new enterprise, White suggested, reminding his new friends that the word *telegraph* in this instance meant "no more than we have agreed to already as to its use."[14]

Just when both parties were about to finalize the terms of the agreement, however, Western Union's attorneys balked at the methods chosen for allocating the last portion (that is, the crucial, controlling portion) of the new firm's stock. The patent litigation was likely to be even more protracted than originally anticipated, they argued, and the danger that a "sharp controversy" in the courts would "expose the weak points in each of the patents" was entirely too high. Novrin Green, of Western Union, admitted to Forbes that such considerations had not occurred to him during the deliberations, and he concluded that little progress in implementing a consolidation would occur "whilst the advises [*sic*] of our respective counsel diverge so widely."[15]

Despite this disappointing setback, the conditions favoring an accommodation continued to exist. All that was needed was an initiative, a new approach that would enable both parties to break the impasse. This breakthrough came from the field (a product in a sense of the tradition and structure that had since the beginning of telephony encouraged local initiatives). In late July 1879, James Ormes, Bell's agent for the southeastern United States, informed the Boston headquarters that he had independently negotiated a settlement with the local Gold and Stock telegraph representatives. The settlement effectively eliminated competition in the seven states served by his franchise.[16] The arrangement, elegant in its simplicity, required the Gold and Stock Company to relinquish the field of voice communication in return for a one-dollar payment on each Bell telephone that the southern agency rented to a customer. The concession was paid directly out of local-company revenues and consequently would not affect National Bell's rental earnings. The southern agency, in turn, would enter into an exclusive agreement with the Western Union and Gold and Stock interests for handling the local collection and distribution of telegraph messages (a service for which it would receive a payment to cover its costs plus a small commission). Forbes was asked to approve the payment schedule, to designate Ormes's seven-state franchise "a neutral ground," and to modify the tenth and eleventh clauses of the standard license contract governing agency relationships with independent carriers in order to permit his local affiliate to deal directly with the connecting telegraph company in developing new business arrangements for handling messages jointly.[17] Ormes himself did not consider this an unreasonable price to pay to protect his position in the market or his investments.

The pact that Ormes had negotiated with the Gold and Stock Company aroused a favorable response in Boston, particularly among those telephone company officials who had been apprehensive about the southern agency's prospects for survival. This large, underfinanced multistate venture had been organized in 1878 under the liberal incorporation laws of New York.[18] Like most of the Bell Company's regional franchises, it was a product of the lenient promotional policies of Hubbard's administration. Since founding his business, Ormes had failed to turn a profit. Only a month before, National Bell's board of directors had rescued the troubled agency from bankruptcy by remitting a small portion of the royalty charged against each telephone assigned to its new Richmond-to-Petersburg toll operation.[19]

Although he was encouraged by the agreement Ormes had negotiated with the Gold and Stock Company, Forbes decided to move cautiously in approving its terms. He was, for instance, inclined to withhold his support until the parties involved had agreed to earmark for National Bell rather than the local licensee the 10 percent commission to be paid by the telegraph company on messages collected from agency subscribers.[20] Forbes now saw considerable potential in this market. He did not want to surrender such an opportunity without some compensation. Moreover, the impression that Bell itself was not the primary spokesman for the local companies in negotiations with the telegraph interests was something Forbes hoped to dispel. All along it had been his intention to project the image of a united front at the negotiating table; a proliferation of independently arranged settlements would have complicated the difficult task of achieving a comprehensive accommodation. To maintain harmony within the ranks, local agents were told only that "something substantial" had, in fact, been accomplished by "moving slowly" and that the Ormes pact was being discussed as a framework for a national settlement. [21]

That the proposed settlement left a number of complicated matters unresolved was reason enough to move carefully in recasting its underlying provisions into a national accord. One troublesome issue not covered in the Ormes pact involved the disposition of American Speaking Telephone Company assets; another dealt with National Bell's potential role in the development of intercity communications services; and a third, with the resolution of conflicting patent claims. All three issues had been finessed under the regional agreement. But whereas Western Union could, without too much difficulty, dismantle and redeploy its own telephones and related equipment outside of the "neutral zone" specified by the Ormes accord, it no longer would be in a position to do so if an agreement were reached on a national scale. There would be no "neutral zone" in any agreement covering the entire country.

It was evident to Forbes that his company would have to absorb the assets of the American Speaking Telephone Company in order to ensure the complete separation of Western Union and National Bell operations. But National Bell was not in a position to purchase these properties outright, so Forbes decided to shift this expense to the licensees. They stood to gain a great deal from the accord, he reasoned. Each exchange could, more conveniently than National Bell, negotiate its own agreement for purchases with the local Gold and Stock Company. [22] The National Bell Company, in concert with Western Union, would formulate a general understanding as to how such properties were to be priced and would establish a timetable for consolidation. Detailed deliberations over the price and condition of the local facilities would be left in the hands of the purchasing licensees.[23]

As the discussions progressed, the arrangements for the transfer of Western Union's local telephone operations fell into three basic categories. In thirty-four noncompetitive exchange areas and in the four areas where Western Union had acquired an interest in the local Bell licensee, the transfer would take place immediately and without complications.[24] In the twenty cities where competition existed, the process of property valuation and reassignment was a bit more elaborate; in these instances telegraph and telephone officials would have to negotiate on a case-by-case basis. There were, as well, thirteen city exchanges in which Western Union held only a partial interest and twelve others in which neither it nor the Gold and Stock Company claimed any direct investment. The two parties merely agreed that at these twenty-five locations they would pursue whatever contractual modifications were necessary to expedite a final resolution of their claims. All Western Union exchange facilities and "interests" were to be transferred to Bell at actual cost, with five exceptions: in Philadelphia, Pittsburgh, San Francisco, Detroit, and New York, Western Union would maintain an equity interest in what was to become a very important group of consolidated exchange companies. Forbes made this concession to ease the pressure on Green from within his own ranks to abandon a settlement and also to ensure that these particular local companies would have adequate sources of capital.

With respect to Bell's future in long-distance services, the two parties wound up reaching an agreement that was ambiguous. The Ormes accord had left the telephone company with what Bell attorneys warned was a very limited opportunity to develop long-distance operations.[25] But actually Green was not completely against opening up the long-distance market to his erstwhile rivals. Primarily interested in protecting his business message monopoly, he was opposed to the development of long-distance telephone service only if it competed with Western Union's traditional operations. Like his predecessor, Green viewed the telephone as an instrument of "personal" rather than "commercial" communications: it would thus naturally occupy the market for two-way, person-to-person conversation that required neither a record of transaction nor an intermediary for purposes of transmission. He was willing to allow National Bell to develop what he believed was the marginal intercity market for social conversation.

In addition, Green indicated that he would not stand in the way of licensee plans to enter the message business in places not served by Western Union dispatch offices; indeed, he hoped that the licensees would act as exclusive, local conduits for messages received over his own long-haul telegraph facilities. For him, such arrangements guaranteed that the only enterprise enjoying a legitimate claim on telephone technology would develop in directions that did not threaten Western Union. Stability would be assured if, as expected, Bell remained in control of the telephone technology and customer preference did not change in some dramatic, unforeseen way. Forbes found these terms acceptable.

The centerpiece of the agreement dealt with the patents. The patents were the fundamental issue underlying the prolonged contest between the two firms; they were the source of Bell's negotiating strength, the legal vehicle for ensuring legitimate monopoly, and the foundation of Bell's contractual arrangements with its local agents and equipment manufacturers. The unwillingness on the part of Forbes and the other Bell executives to compromise their firm's claims before the public and in the courts had, in effect, sentenced the earlier efforts at consolidation to failure. With National Bell and Western Union now engaged in fashioning an industrywide alliance, however, it was in Western Union's interest to reinforce as much as possible Bell's claims over the telephone technology. Neither firm wanted potential competitors to develop telephone exchange systems. Therefore, Green agreed to turn over to Forbes all of his firm's patents and inventions in the field of telephony.

A steep price was paid by National Bell to regain its monopoly over telephone technology. Under the agreement, Western Union was awarded a 20 percent royalty on the gross earnings derived from National Bell's telephone rental business (less the standard commission that the company routinely granted its licensees).[26] Payment was to continue for the life of the patents. Green also asked for and received a 20 percent share of any net profit that National Bell earned from exports. National Bell officials also accepted restrictions on their ability to cut prices under normal conditions (mostly to reassure Western Union that its share of total telephone revenues would never fall below one dollar per instrument). In return, Western Union transferred all of its telephone patents, including those dealing with switching equipment, to the Bell interests. Green even committed his company to defending his erstwhile rival's patent interests by agreeing to share one-fifth of the legal costs associated with impending or future infringement suits. The amount that Bell actually paid for this settlement might have been even greater had not the company's attorneys carefully maneuvered to exclude certain revenue sources from such claims. Still, the royalty that Western Union received would come to represent a large slice of Bell's profits.[27]

To Bell officials, the restoration of their firm's dominant market position was worth the price. To Western Union the surrender of what its management had almost always viewed as a marginal business seemed a reasonable sacrifice for the promise of industrial peace and the end of Gould's threat. Accordingly, the two parties signed the agreement on 10 November 1879, bringing to a close the first competitive chapter in Bell's history.

The competition had dramatically quickened the pace of expansion. As the two firms had raced to occupy the field, exchanges had been built without a great deal of thought to the business's long-run technological or financial requirements. In the years that followed the Agreement, Bell officials would have to deal with the many problems left behind by this competition.

Armed with the agreement, and able to attract new investors, Bell's resources for dealing with these problems would be considerably enhanced. Vail and his coworkers would make good use of these expanded resources to improve and eventually integrate the operations of the various telephone concerns. As yet, however, there was no single Bell system, no national network. In the years following 1879 the underpinnings for such a system would be built.

CHAPTER 5 🖋️

Building the Foundations for the Modern Bell System

WITH THE WESTERN UNION ACCORD IN PLACE, National Bell's officers had time to reflect on the position their firm had attained in the marketplace, its strengths, and the challenges it would probably face as it entered the new era of patent monopoly. While it would still be necessary for the Bell interests to defend the patents, Forbes and his colleagues could be reasonably confident that there was no competitive threat on the horizon comparable to the giant telegraph company. Instead, management's immediate concerns would be of an internal and strategic nature: the process of administrative consolidation of the exchange business, temporarily postponed during the feverish rush into the field, would have to begin soon; providing greater financial support for licensees would at least have to be considered; and the technology and finances of establishing the interexchange business would impose considerable demands on the company. The only decisive external problem that the firm would face—a problem that could not have been anticipated in 1880—would involve Bell's first major encounter with the government and would be precipitated by the firm's need for additional capital as demand for telephone service soared in the prosperous 1880s.

When Theodore Vail compiled the figures needed by Forbes for his 1880 report to the stockholders, the totals were impressive. At last count, nearly 61,000 telephones had been installed—38,000 in the last year alone. Service was now available in 998 American cities and towns; all but 14 of the nation's largest municipalities (over 10,000 inhabitants) boasted an exchange. Demand for central office services had exceeded even the most sanguine predictions of the system's early proponents, and the talk about market saturation (in New England) now appeared to have been far off the mark. In a few cities 1 out of every 25 residents had leased telephones, a figure four times greater than what had once been considered the maximum demand.[1]

Telephone demand had surged during the Bell–Western Union competition, and what was even more remarkable, it showed no signs of abating. In the first two months of 1880 the Williams shop had produced ten thousand phones. At this pace, National Bell would exceed the past year's entire production run early in the third quarter. These figures were all the more striking in light of Bell's "understanding" with Western Union suspending, in mid-1879, the licensing of new agents until a general framework for a settlement had been negotiated. Vail had not resumed enlisting new agents until mid-December, which suggested that growth in demand was being largely sustained by the intensive development of existing markets, not by expansion into new territories.[2]

Exchange development through 1879 had built a strong foundation for earnings in the 1880s. "Old business" alone, Vail guessed, would account for about $270,000 in revenues in 1880, even after allowances were made for equipment replacements, for Western Union's 20 percent commission, and for instruments "out at cut rates."[3] Moreover, unit profit margins were expected to improve as the company implemented its plans for a selective but steady price increase on leased equipment. Bell's Chicago and New York operations would continue to be a drain on corporate resources, but now their requirements could be more deliberately tailored to available capital instead of to competitive exigencies. Overall, the prospects for a substantial—and badly needed—increase in the company's cash flow seemed bright.

Increasing demand seemed to promise a bright future for the National Bell Company, but this increasing demand, as well as the new complexity and scale of telephone operations, was also creating some of the firm's most vexing technical and administrative problems. Matching manufacturing capacity with the explosion in demand for telephones and related equipment constituted one major concern; improving the quality of production, another. In the field, the licensees needed assistance in surmounting what appeared to be a stubborn economic barrier to the extension of their central office services; the cost of providing exchange connections was rising more rapidly than the revenues collected under the prevailing flat rate schedule—a phenomenon that would eventually require a fresh approach to both the design of switchboards and the pricing of services.[4] The expense of constructing exchanges had exceeded, frequently by several times, the original estimates, and the existing systems, installed during the rush to occupy the field, suffered from defective "internal workings." It now appeared that many of the existing setups would have to be replaced as soon as feasible and at considerable cost.[5]

The solution to problems of this nature would require considerable expansion in National Bell's technical and administrative resources. But undoubtedly weighing most heavily in the minds of Bell executives were the financial ramifications of these new demands and opportunities. In late 1879 Forbes had appointed a committee to review the company's long-term prospects and to submit specific recommendations that would serve as an agenda for extended deliberations on the company's future business strategy.[6] The report underscored the increasing array of commercial opportunities arising out of advances in telephone transmission and central office switching technology and suggested that expansion into these new areas would provide the most promising means of assuring the long-term vitality of the business. But a move in this direction almost certainly meant a change in National Bell's relationship with the licensees, since it would be entering operations that had from the beginning been entrusted to its agents.

In the short run, such changes would come slowly. The pace and breadth of the parent company's move into operations would depend largely on its capacity to raise capital and develop its own managerial resources; it would of necessity be carefully calibrated. Existing financial circumstances, the committee reported, precluded any immediate venture into new, capital-intensive businesses. The

committee concluded that it was "not wise to undertake the ownership and control of the entire system of district exchanges [since] this
would require a very large capital and a group of officers such as
cannot for many months be selected and educated to the business."
A similar conclusion was reached with respect to the company's
potential role in the market for interexchange toll service. "As far as
may seem wise," National Bell should "be prepared to build and
own" toll-line facilities, but given the firm's limited resources, it was
probably "best to leave the ownership of these . . . temporarily with
the local companies."[7] Not until National Bell's capital had been
greatly extended would such investments become feasible.

These recommendations, along with the agreement negotiated earlier with Western Union, set the stage for plans, announced in early 1880, for recapitalizing the business at $10–$15
million and for reincorporating as the American Bell Telephone
Company.[8] It was certainly an auspicious moment for refinancing the
business. The price of National Bell's stock on the open market had
soared to almost $600 per share, nearly six times higher than what it
had been as recently as the summer of 1879.[9] Investors now had
confidence in the enterprise and its management. But as Bell officials
discovered when they petitioned the Massachusetts legislature for a
new charter, not everyone shared their enthusiasm for a secure and
prosperous future as an entrenched patent monopoly.

Earlier the company had been warned that as a monopoly it
had a less favorable public image than it had once enjoyed as a
struggling, competitive enterprise. J. J. Storrow, Bell's leading
attorney and legislative representative, had alerted officials to the
considerable concern that the emerging telephone monopoly had
produced within the state legislature. He advised his colleagues
against seeking unnecessary or extraordinary privileges in petition-
ing for a new charter; in a memo circulated in early January 1880,
Storrow took the position that "we merely want a legal existence
corresponding to our growth . . . we do not want any rights different
substantially from those given by existing laws." He urged the Forbes
administration to go public, within the bounds of prudence, with its
plans for the business, to justify corporate growth on grounds that
made sense in light of the peculiar economic characteristics of the
industry. "When it is remembered that the capital asked for will not

enable us to put up 1/100 of the lines in use within a year it is apparent that our profit will come from extending the use of the instrument . . . and not from monopolizing lines," he told Forbes.

> If it be suggested that we furnish [telephone service] to everybody who applies, the answer is this; certainly we intend, and as shown, our interest will compel us to furnish for the public to the fullest extent—but suppose we want to have an exchange in Worcester, we must induce some one to put $50,000 into plant, which will be of little value if some one else can do the same thing. Moreover, the value of an exchange to the public consists in connecting every-body with one exchange. . . . Our interest and the good of the public therefore, requires us often to grant exclusive territorial rights coupled with an obligation to be diligent in extending the use of telephones by furnishing them to all.[10]

Storrow recommended that all new licenses include "a condition that the party, agent or whatever . . . shall do his best to meet *every reasonable demand* of the public at *reasonable* prices." He hoped that such an explicit statement of intentions would temper the fears of those legislators who seemed bent on regulating the firm.[11]

Before anything of this sort could be done, however, National Bell's political opponents made their first move. In late January the Senate Committee on Mercantile Affairs reached a decision on the company's plans for reorganization; it cut American Bell's proposed capitalization in half, prohibited the use of such funds for the purchase of new patents, and placed telephone rates under the jurisdiction of the state's Board of Railroad Commissions.[12] Dismissed as irrelevant were Storrow's protests that telephone ser-vice constituted a form of "personal communications . . . essentially private in nature" that was not liable to regulation under existing precedent.[13] The company's policy of assigning agency franchises on an exclusive basis became the target of a great deal of criticism; for the moment, at least, the lawmakers appeared determined to limit through regulation American Bell's capacity to charge what they considered exorbitant rates.

A number of influential newspaper publishers had joined the battle. As a group, they regarded the Western Union accord as an unethical, if not precisely illegal, barrier to lower charges for trans-mitting messages. In the past, National Bell had been regarded as a competitive check on the telegraph monopoly. Those hoping for some relief had been greatly disappointed when provisions restricting Bell's ability to carry such messages had been included in the agreement. Forbes came to regard their opposition as the most serious threat to his efforts to usher through the legislature a reasonably "untrammeled" charter. "If this matter is agitated at present we are likely to have our charters loaded with amendments regarding public messages and news dispatches. . . . Any laws passed about this business will be examined and perhaps copied in other states as telephone questions come up."[14]

After discussions with Forbes, Western Union officials grudgingly agreed to a selective relaxation of the restrictions governing the transmission of news accounts over telephone lines. Shortly thereafter, the publishers abandoned their opposition to the company's new petition. But state lawmakers remained disturbed over the prospects of American Bell's extending its control over its licensees (even though the legislators had by now acknowledged the validity of the technical and economic arguments in favor of local monopoly). The charter that emerged from the second round of hearings therefore prohibited American Bell from securing more than a 30 percent stock interest in licensees located in the state of Massachusetts. The provision for rate regulation, however, was dropped from the final measure, much to Forbes's relief. On 12 March 1880 the lower house passed the bill of incorporation submitted by its committee, and a week later, the state senate enacted similar legislation. The bill was signed into law on 2 April 1880, and American Bell opened for business about two weeks later.[15]

The new corporation was empowered to construct, own, and maintain "public and private lines and district exchanges" and to "become a stockholder in or become interested with other corporations hereafter organized for like purposes, or already established for the transaction of the telephone business under its patents."[16] For these purposes, the act allowed management to recapitalize the undertaking at $10 million (not the $15 million originally requested). While the new charter was framed in a manner that largely fulfilled management's objectives, the Bell interests had not escaped the legislative process entirely unscathed: their power to own stock in their Massachusetts licensees was limited, and what was perhaps more ominous, the subject of rate regulation had received full, public airing. For the moment, the company's political problem had abated, but it should have been clear to all concerned that proposals to regulate rates and services were destined to arise again as the telephone became a more important factor in American urban and commercial life.

Inside the firm, Vail had busied himself with the task of bringing the company's first-generation licensing agreements into line with the new technological and economic conditions in the industry. During the closing months of 1879 he had prepared new and revised contracts. No longer did the exchange agreements (the form 109 series) confer large territories on the agents; generally, franchise boundaries were limited to a single municipality or town.[17] In the view of Vail and Forbes, this change would enable Bell agents to more easily match their administrative and capital resources with the expanding demands of central office operations. Expectations about local performance were frequently spelled out in amendments specifying a timetable for the construction of central office facilities; numerical targets for subscribers and rental revenue growth were usually provided.[18] A new emphasis was placed on tightening the company's licensing arrangements and accelerating development in the field. In the spring of 1880, contracts were issued for the new branch line, toll, and messenger service businesses which would extend the Bell interests into new markets in the following decade.

Most of the contracts issued at this time were of no more than five years' duration and (like those they replaced) contained provisions granting Bell the right to acquire agency properties at cost at the time of expiration.[19] Bell management clearly was looking down the road to the time when advances in transmission technology might force yet another reconfiguration of licensee operations. The licensees, of course, had hoped for a more permanent arrangement that would have improved the climate for investment in their local companies. But with the exigencies of intense competition past, the Bostonians in charge of the revitalized Bell enterprise could afford to postpone decisions that might otherwise have locked them into a structure of industry relations and a menu of obligations that they would later regret. Time was on their side. Before a convention of exchange owners unhappy with the length of term of the new contracts, Oscar Madden explained: "The business is so new, so different from any other, it is every day presenting new phases; changes in the contracts of today may be as necessary as the changes in the past. To sum up, it was thought best, and in the interest of all concerned, that the business should have the benefit of more growth and experience before 'long term' or 'permanent' contracts were made."[20]

Of all the potential developments that seemed likely to force further changes in the way local telephone systems were organized and managed, the toll or interchange business was the strongest candidate. At this time the toll business was technically in a primitive state. Voice transmission was not feasible over great distances, and interconnection of exchanges was extremely difficult owing to the variety of central office facilities and operations. Totally new as a commercial venture and of very limited range in practice, toll operations connecting neighboring exchange offices had nevertheless inspired a profound interest among a small circle of Bell officials. They anticipated significant commercial opportunities in this field and chose to wait until the situation was clarified before deciding what role the licensees would fulfill in developing such systems. It "may prove undesirable," Madden said, "to issue licenses for a single city or town, as we are now doing, [since] the development of the extraterritorial line business will bring about no inconsiderable change in the relations of exchanges to each other. Isolated now, they will become, in many cases, unified in groups, having one common interest, and seeking one common management. I look to see consolidations all over the country, such as have already begun in the East."[21]

In the meantime, Bell encouraged the licensees to cooperate as the new toll business was developed and thus to exploit the crucial economies of sharing. As much as possible, extraterritorial services were to employ the existing exchange plant as the primary conduit to local subscribers. Central office operators were to provide local connections. Among the new provisions written into the revised exchange contracts were terms obligating local affiliates to furnish incoming toll messages with "the proper switch board and other connections" and provisions establishing the ground rules and charges for sharing right-of-way privileges and connecting agency poles and lines.[22] Local telephone companies were also assigned responsibility for the collection of subscriber toll charges to minimize the expense and simplify the administration of the new extraterritorial operations. The additional costs that exchange agencies incurred in the performance of these duties were to be covered by a modest "originating commission," paid by toll-line operators, which usually amounted to 15 percent of the toll revenue generated by subscribers. American Bell, for its part, received a 25 percent commission on the toll business, a sum compensating it for what its management described as the loss of instrument royalties in this valuable new area of business.

Largely unchanged under the revised contracts was the authority exercised by the licensees over decisions about the rate and character of exchange plant development and the deployment of central office operating personnel. In time this became a problem. Those interests financing and managing toll facilities during this early stage of the industry's development would find frustrating their dependence on "affiliated" exchanges for district connections. In general, the local companies favored the commercial expansion of their own endeavors over toll businesses, which seemingly were much less certain of success and much more modest in earnings potential. Toll operations were pushed to the back burner while the licensees dealt with their own pressing exchange requirements. In the end, a number of the industry's most ambitious extraterritorial undertakings would collapse, prompting American Bell policymakers to look for new ways to organize and manage this important area of the business.

In 1880, however, the toll business was full of promise. During the previous year construction had been completed on the nation's longest extraterritorial system, a twenty-eight-mile line between Boston and the prosperous textile community of Lowell, Massachusetts.[23] Charles Williams built the line for the Pioneer Telephone Company, an operation financed by prominent Lowell businessmen who also held an interest in their local exchange business. The toll system was in commercial operation by late 1879. Its very name suggested the novelty of the undertaking; the same could be said of the cooperative arrangements that bound connecting local agencies and the Pioneer Company together in what was presumably a single, unified system.

From the outset, the line was plagued with transmission interruptions and interference. Some of these problems were solved with the help of Thomas Lockwood, of American Bell's technical staff, who reconfigured the Pioneer Company's two grounded iron wires into a higher-grade, single "metallic circuit" system.[24] But the line continued to require a great deal of attention to sustain even the most modest levels of transmission quality. As maintenance and repair costs ballooned, connecting local telephone companies began to curtail their efforts to keep the system in good running order.

The biggest headache arose in Boston, where the problems of interference were compounded by that city's expanding street railway system and electrical utility businesses. Shortly after the line opened for commercial use, American Bell began to receive complaints from Pioneer Company officials, who accused the connecting Boston Dispatch Company of neglecting to repair its section of the line and of mishandling incoming messages. William Witcome, responding for the Boston exchange, angrily objected to "being forced to transact the business of Exterritory [sic] lines," implying that this business, along with its attendant obligations, represented an unsolicited intrusion on his agency's prerogatives, as well as an unwelcome expense. The Forbes administration tried to negotiate a compromise but failed. Barely a year after launching this promising experiment in long-distance transmission, American Bell had to assume full responsibility for the management of the line.[25]

American Bell's officers tried to find a way to make the system work. For instance, they revised the schedule of compensation, raising the commissions given to the connecting exchanges from 15 percent to 25 percent. A strong case had been made by both the Lowell and Boston exchanges that the original formula had not covered their costs, and these arguments were accepted.[26] But this compromise did not really get to the heart of the problem. The size of the original commissions did not completely explain the line's problems in Boston, for at the Lowell end of the system, under the same schedule of charges, no difficulties had been reported. There, where investors held an interest in both the local exchange and the toll business, the local company had been willing to put up with the trouble and expense stemming from extraterritorial operations. This suggested that consolidation of the exchange and toll system managements—and their combined ownership—might represent the only reasonable means for integrating these two essential components of the emerging system.

Had the Pioneer Company been the only major failure during these formative years of the toll system, the American Bell Company might not have contemplated such a decisive shift in business strategy. But there were other disappointments, some of them involving larger, more complex systems. All appeared to have one problem in common: great difficulty in obtaining suitable connections into the local exchanges. In 1883 a one-hundred-mile line built between Chicago and Milwaukee was abandoned after the officers of the businesses involved were unable to reach an understanding on either the technical specifications for transmission over such a distance or the best methods for handling incoming messages at the exchange agency level.[27] Indeed, Charles Fay, of the Chicago exchange, sank the entire venture by refusing to turn his central office into what he disparagingly described as "a relay office" for toll traffic arriving from Milwaukee.[28] Similar controversies arose over a line built between Kansas City and Leavenworth, and between the operators of a toll system in Missouri and officials of the St. Joseph exchange.[29]

To Bell officials, the most shocking of these incidents was the collapse of the Interstate Telephone Company. This ambitious enterprise had been founded by a consortium of prominent New England capitalists and politicians in the summer of 1880.[30] Although its charter described it rather grandly as a company that would eventually link the major centers of commerce and industry along the Atlantic Coast with their counterparts in the Midwest, its first endeavor was of modest proportions: a line between Boston and Providence, Rhode Island (where many of its original investors lived). With a liberal charter from the state of New York in hand and about one hundred thousand dollars in the treasury, Interstate Company officials began construction in July 1880 on the first link of what they hoped would one day become an extensive, interregional toll system.

Both Watson and Vail closely monitored the line's progress as it slowly wound through eight New England towns.[31] Built on sturdy, forty-foot chestnut poles each carrying seven number-nine galvanized iron wires, the Interstate line was clearly the flagship of long-distance operations. To the dismay of Vail and other American Bell officers, however, Interstate's board abruptly voted to dissolve the business only eleven months after launching it.[32] The losses suffered in the first half-year of operation had been very discouraging: revenues of only $669.15 to cover expenses of $1,555.42. "Defects" in the line had curtailed service and dampened subscriber demand. Only two of the seven "temporary" wires originally requested by Interstate had been extended to New Hyde Park by the Boston exchange.[33] Crossed circuits at the Boston end of the line had further frustrated the company's attempts to get connections into that city. Interstate's general manager had felt powerless to correct many of these problems, particularly those encountered in and around Boston. Desperate for a resolution, he had pleaded with American Bell to intervene on his behalf, claiming that he had "no jurisdiction whatever" over the wires in question.[34]

With the prospects of additional failures on the horizon, the Forbes administration moved decisively to correct what it viewed as primarily a problem of organization. A loose combination of interests—held together by a variety of licensing arrangements—had been tried and had failed. Following a course common to most industries facing problems of coordination and control in these years, Bell pressed toward a tighter form of combination, one that would enable the new, consolidated businesses to manage such disagreements internally. A decision was made in late 1881 to begin consolidating certain toll and local exchange operations in areas where these combinations could be formed around a coherent and identifiable community of interests. The new policy was announced to shareholders in Bell's annual report for 1883:

> The tendency toward consolidation of telephone companies [was], for the most part, in the interest of economical and convenient handling of the business. The connection of many towns together, causing large territories to assume the character of great telephone exchanges, made it of importance to bring as large areas as possible under one management to insure simple and convenient ar-

rangements for furnishing rapid intercommunications. As methods are devised for making the telephone commercially useful over long lines, the advantages of this centralization of management will be more apparent, as well as the importance to the public of having the business done in large territories under one responsible head, with far reaching connections throughout the whole country. To make this service the highest value to the people will be complicated enough under one control.[35]

Instead of rushing to impose the new structure on the entire industry, American Bell embarked on the process of centralization in a relatively cautious manner. Change was incremental and tailored to a considerable degree to fit local circumstances. This, in fact, would be the pattern of administrative change evident throughout most of the company's history. Care was taken to preserve those longstanding traditions of entrepreneurship and managerial autonomy that officials regarded as so crucial to their company's past success. "In the interest of convenient and economical management of the business, consolidation was not [to be] encouraged to an extent that would leave these companies entirely in the ownership of persons who [were] not residents in the territory where the business [was] carried on," Forbes explained. The new realignment would "keep local capital and influence interested in the business as far as possible" to help maintain Bell's roots within the communities its licensees served.[36] Such combinations, shareholders and public officials were assured, did not prefigure any further expansion in American Bell's control; they were designed to eliminate organizational impediments to the establishment of an efficient intercity system of communication.[37]

The new structure heralded another important step in the conceptual and organizational development of the business. Moreover, it also represented a crucial shift in the kind of factors shaping the administration of the business. During the first two phases of Bell's development, franchise boundaries had been shaped and revised in accordance with local financial and entrepreneurial capabilities and with the particular capital requirements of either the private-line or the exchange operations. Now a new logic of system formation, managerial centralization, and operational integration was taking hold. The operational and technical demands of the network were pressing certain areas of the enterprise toward a higher level of organizational integration. The years that Forbes had spent piecing together small New England railroads to form larger regional transportation networks no doubt influenced his outlook regarding the reorganization of the telephone business.[38] Vail, too, having once managed the Railway Mail Service, instinctively attributed efficiencies to large, centrally coordinated systems.

The process of telephone consolidation took place first in the East, the most familiar and concentrated market for toll-line expansion. Beginning in 1882, operations in New York State were regrouped into several large systems, among them the Hudson River Telephone Company, a combination of ten agencies situated between Albany and Fishkill, New York; the Central New York Telephone and Telegraph Company, a business refashioned out of the American District Telegraph Company of Utica, the Merkimer and Mohawk Valley Company, and the Ogdensburgh and Watertown agencies; and the New York and Pennsylvania Telephone and Telegraph Company, an enterprise with interests in thirteen exchanges spanning the territory between the shores of Lake Erie and Ithaca, New York.[39]

Along similar lines, in 1883 the New England Telephone and Telegraph Company added twelve new exchanges in the states of Maine, New Hampshire, Vermont, and Massachusetts to its expanding family of properties. The process was repeated elsewhere: in Washington, D.C., and in Baltimore, Maryland, where exchange operations were regrouped under the Chesapeake and Potomac Company of New York; and in thirty counties of Pennsylvania and one in New Jersey, where the Central Pennsylvania Telephone and Supply Company (itself an earlier combination of several exchanges) was consolidated with five other interests (including properties once belonging to Gardiner Hubbard).[40]

The phenomenon of agency and exchange consolidation spread as well into the less populous heartland of America. Eleven agencies covering the states of Illinois, Ohio, and Indiana and a portion of Iowa were merged in 1883 into an amalgamation that became known as the Central Union Telephone Company.[41] The same change took place in Missouri, Kansas, and Nebraska in

January; in the remaining portions of Iowa in June; in Arkansas and Texas in July; and in Idaho, Montana, Utah, and Wyoming in April of 1883. In each case a large, multistate operation was fashioned out of what had once been small isolated businesses.[42] By 1884 the industry had undergone a fundamental restructuring, one that laid the groundwork for a more unified approach to the planning and administration of the emerging Bell toll systems. This process would continue for many decades to come, but few events would have so decisive a long-term impact on the business as the 1881–84 reorganization.

As these dramatic changes were taking place, the commercial development of the toll business was rapidly accelerating. In 1883, 10,616 miles of new wires were added to the existing system, more than tripling the capacity of Bell's extraterritorial plant. Toll-circuit mileage doubled again in 1883 and grew thereafter at a steady rate of between 35,000 and 36,000 miles a year.[43] Ironically, when the demand for instruments tapered off unexpectedly in 1883—temporarily ending three years of remarkable expansion in telephone production—company officials attributed it in part to a shift in local attention and resources to development of the toll network.[44] For the licensees, exchange service was no longer the only major commercial opportunity demanding attention and investment.

Accompanying these dramatic operational realignments at the local-company level were important changes in the scope and administration of American Bell's own business. Eighteen eighty-one was the year that Forbes chose to move on several of the long-term, strategic recommendations that the committee had tendered him in late 1879. His first step was to consolidate his company's position in the manufacturing and supply end of the business. In July 1881 a 40 percent interest was acquired in the Western Electric Manufacturing Company, the nation's leading supplier of electrical equipment (and until then Western Union's primary source of communications gear). Four months later, a new corporation, representing the combined properties of Western Electric, the Gilliland Electric Manufacturing Company, and the Charles Williams shop, became part of American Bell's expanding set of interests. An agreement designating the new enterprise as the exclusive supplier of patented telephone equipment for the licensees, signed in

February 1882, ended the brief period during which Bell had relied on five independent manufacturers for the production of call bells, switchboards, and other ancillary telephone devices.[45] By the close of 1882 American Bell could count as part of its enterprise manufacturing plants in Chicago, Boston, New York, and Indianapolis, and its management notified the shareholders that it harbored high expectations of making its new manufacturing arm "an important and valuable part of our business."[46]

At about the same time that the Forbes administration began piecing together these scattered production facilities, the company set in motion ambitious plans to obtain "a permanent vested interest in the telephone business independent of its royalties upon telephones."[47] The centerpiece of this plan was an offer by American Bell to replace its five-year licenses with new, permanent contracts in exchange for a stake in the reconstituted telephone companies. Usually no more than a 35 percent interest was required. Where special royalty concessions or particularly lucrative territorial assignments came into play, as much as 50 percent was demanded.[48] Dividends on such "franchise" equity were to be suspended, at least until the original five-year contracts had expired. From American Bell's point of view, this constituted a reasonable exchange. The stock represented a claim on the future profits of the exchange business and gave the licensees some assurance that parent company management had decided against entering this field itself upon the expiration of the existing five-year contracts.[49] The licensees, though perhaps not enthusiastic about the terms of the agreement, could find solace in the fact that their contracts were permanent and would finally enable them to plan the development of their businesses on a long-term basis.[50]

The acquisition of an equity position in the licensees and in Western Electric was part of the same corporate process and strategy: the integration of Bell's affiliated properties. The ties with Western Electric were a form of vertical integration. Horizontal consolidation at the local-company level represented a complimentary trend, the first crucial step in the direction of systems integration. Like many of the large corporations that were to emerge during these years or shortly thereafter, American Bell was extending its control in both directions. Vertical integration immediately brought with it the kind of managerial oversight the company had long sought in the realm of manufacturing and distribution of its patented devices. However, the same could not be said at this time with respect to the company's new interest in the licensees. In that case the customary inclination to allow for a relatively high degree of local autonomy in building and operating these businesses continued to hold sway.

The financial integration of the enterprises, however, did bring with it new incentives for the extension of centralized managerial oversight in the areas of local earnings and operational performance. With the sources of American Bell's revenue now expanded beyond the instrument-rental business into telephone operations, the company began to pay more attention to the service and financial performance of the licensees. Watson's small Electrical and Patent Department, organized in 1880 primarily to evaluate the patents and devices developed by independent inventors, began to turn out publications designed to instruct agents "as to the best methods of building and operating exchanges and other telephone systems."[51] Madden's traveling agents, who in the past had concentrated almost exclusively on the commercial aspects of licensee affairs, were now kept "well posted" on the new products and services coming out of the company's experimental laboratory.[52] After gaining some "early practical experience" in the field—what he only half-facetiously referred to as the "Valley of the Shadows of Death"—Doolittle was given more assignments in the field to help compensate for the lack of technical expertise within the ranks of the regular traveling agents.[53]

One way in which the company improved its "oversight over licensee affairs" was through changes in the scope and detail of its basic accounting system. With such modifications, Forbes sought "as far as can be obtained without absolute ownership, an exact comparison of results."[54] In 1884, when his administration unveiled a newly "shaped" American Bell to its shareholders, he assured them that the enterprise had brought its several parts under tighter control and that "with proper watchfulness, we may expect steady growth and improvement in the character of our business in all its branches."[55]

Driven by the need for more accurate information concerning all aspects of the business, the company more than doubled its headquarters staff between 1882 and 1885. Twenty-nine employees, nearly half the new total, were now assigned to licensee affairs, general bookkeeping and records, and the administration of instrument-rental accounts. But the greatest growth occurred in the technical staff, where, in 1885, twenty employees were divided between the Electrical and Patent Department and the recently organized Mechanical and Testing Department.[56] The former organization had been split into two specialized units in December 1883 owing to the formation of a new "experimental shop" that separated from Bell's customary functions the development work of Emile Berliner and W. W. Jacques and the patent appraisal activities of Thomas D. Lockwood. Routine "experimental work" relating to circuit design and equipment inspection, once the province of the Electrical and Patent Department, had been transferred to the Mechanical and Testing Department, now under the direction of Hammond V. Hayes. Between 1882 and 1885 the staff that Watson had managed as Bell's general inspector of instruments increased from two to twelve as Western Electric rapidly increased its production. All of these units reported directly to the general manager's office.[57]

Vail's staff could now gather from the field a vast store of general information concerning licensee operations, rates, earnings, and markets. Thirty-five companies were surveyed by questionnaire on these and related matters, and the results gave American Bell officials their first comprehensive picture of what was happening outside of Boston.[58] The licensees were also queried on such subjects as the possibility of reducing residential rates, the quality and different classes of service offered to the public, and the elasticity of demand for telephone instruments and services.

An even more elaborate effort to obtain timely, useful, and detailed information on the licensee operations began in 1885. Bell realigned around a specific cluster of telephone companies the traveling assignments of its agents in the field; this was done on the premise that through their acquaintance with the same operations, they would gain a "minute and particular knowledge" of the business. Agents also began providing the home office with reports documenting trends in state and municipal legislation (including copies of pending bills dealing with telephone regulation) and prepared on a routine basis accounts of the whereabouts and operations of "infringing companies." After 1885, traveling agents were appointed to the boards or executive committees of several of the local telephone companies that they regularly visited. Moreover, they were now encouraged by the home office to acquire enough practical knowledge and experience in managing exchange operations to enable them to take charge in the event of an emergency.[59]

The expansion and reorganization of American Bell's headquarters staff and the extension of its oversight to key areas of the business were major steps in the administrative consolidation of the system's horizontal component. The Forbes administration now monitored the performance of local companies in which it held a substantial investment. No longer was the parent company's welfare dependent solely on royalty earnings; its substantial and growing stake in the toll and exchange businesses had changed dramatically the composition of its income. By 1884 American Bell was a far different entity than the loosely coordinated set of interests of 1880, which had been unified, in large measure, by a set of temporary licenses and a strong patent position in most phases of telephony.

CHAPTER 6 🖋

The American
Telephone and
Telegraph Company:
A New Beginning

BY THE MID-1880s THERE WAS GOOD REASON to believe
that Bell officials had surmounted many of the organizational and
economic difficulties that had plagued the development of regional
toll systems. The recently consolidated operating companies now
commanded the resources needed to build their intercity operations,
and Forbes and Vail appeared confident that the carefully crafted
realignment of licensee properties would quickly eliminate the
conflicts of interest and technical obstacles that had originally
blocked the commercialization of long-distance services. By acquiring stock in the operating companies and by imposing a small
measure of administrative control over their activities, American
Bell had apparently gained the leverage it needed to ensure cooperation among those involved in building and operating these systems.

In early 1885, however, the limits of their new approach to
organizing the regional toll business became evident. At that time,
the management of the Southern New England Telephone Company (SNET) announced it was abandoning its two-hundred-mile
experimental toll line between Boston and New York City. The line
had been a solid technical success; problems of transmission over this

Edward J. Hall, Jr., the first general manager for American Bell's long-distance subsidiary, AT&T. Hall was recruited into AT&T by Vail and Forbes, who viewed this innovative vice president of the Buffalo Telephone Exchange as an experienced leader within the new industry. In the fall of 1887, Hall was appointed vice president of AT&T, a position he occupied until his death in 1914. His many contributions to the field included the establishment of the first, feasible system for charging subscribers on the basis of telephone usage rather than on a schedule of flat rates and several important insights into the most effective way to organize the overall business.

1889

The original Bell seal, first used to distinguish AT&T's long-distance service.

distance had, for the most part, been solved. But disappointed SNET officials found the expense and capital requirements of the long-distance business much higher than originally anticipated. The line was a commercial failure. Its earnings were marginal at best. By the end of 1884, the spirit of entrepreneurial adventure that had driven SNET officials and their investors to the frontiers of telephony had given way to a more pragmatic outlook. With the demand for local service expanding, the company turned away from the long-distance business, preferring to concentrate its attention and resources on developing its more profitable exchange and short-haul toll services.[1]

Within American Bell, SNET's decision prompted management to take a more active role in long-distance operations. Neither Vail nor Forbes could justify an about-face on their commitment to the field of intercity telephony, especially since the technical feasibility of long-distance transmission was no longer in doubt. Moreover, Vail, who was beginning to chafe under Forbes's diligent but conservative leadership, viewed the prospects of American Bell's entrance into long-distance communications as a unique opportunity to get out from under Forbes's thumb and to create an administrative domain of his own. In early 1885, with the blessings of his Boston colleagues, Vail resigned his position at the American Bell Company and began assembling a small team to plan the organization of a wholly owned long-distance subsidiary. As his right-hand man and general manager he recruited Edward J. Hall, Jr., of the Buffalo exchange, an agent who had distinguished himself in the field of local telephone service and who would later come to be known as the father of the modern message-rate system.[2] Angus S. Hibbard, at the time an employee of the Wisconsin Telephone Company, was chosen as the new company's general superintendent. Over the next few months these three men worked together to draft a charter and formulate a viable commercial strategy for their enterprise.

Vail instructed Hall to incorporate the new long-distance subsidiary, named the American Telephone and Telegraph Company (AT&T), in New York State, which offered a far less restrictive political environment than did Massachusetts. The charter was to include rights to increase AT&T's financial resources to "an unlimited amount" as the business grew (initially the plans called for a capitalization of no more than one hundred thousand dollars).

"Make the powers of this Company to build, buy, own, operate, lease, etc. . . . lines extending from any city in the state, to each and every other city, and also from any city in the state and every city in the U.S., Canada and Mexico, and to be connected by cable with the rest of the known world," Vail told Hall, adding that the charter should be as extensive as possible "for if it is to grow into a large Company, we shall want unlimited rights."[3]

Incorporated in this expansive manner, the new subsidiary would not suffer the same financial restraints that American Bell had suffered under Massachusetts law. But there was, nevertheless, a serious question as to the economic viability of its operations. Long-distance telephony was a new and unproven field of endeavor on the leading edge of communications technology. Unlike the many ventures in which American Bell had invested to date, AT&T was a business that would have to be organized from the ground up. It was not a well-established operation like Western Electric or the local telephone companies.

Perhaps even more challenging were the economic characteristics of the business. Early on, Hall determined that a private-line, point-to-point long-distance system, though simple to organize and administer, probably would not generate enough traffic to justify American Bell's investment in it.[4] Instead, he believed that AT&T's success as a commercial venture depended on its capacity to use all existing exchange and toll facilities belonging to Bell licensees as "feeders" into its own intercity network. Integration in this manner would minimize AT&T's capital expenditures and maximize its coverage of the national market.

Hall thus hoped to develop between AT&T and the licensees a structure of cooperative arrangements under which AT&T's expenditures would not exceed 10 percent of the total long-distance revenues.[5] For a nominal fee, local telephone companies would provide on-site engineering support, right-of-way privileges, and space on their telephone poles; they would also collect tolls from subscribers who used Bell's long-distance service. It would probably be necessary to provide temporary subsidies earmarked for the construction of the local "feeder" lines, or rebates on the extraterritorial commission that American Bell earned on facilities rebuilt with the requirements of long-distance transmission in mind, Hall reasoned.

But these obligations were not thought to be so large as to reduce dramatically the profits of the business. Indeed, they seemed to be a relatively modest price to pay for the large savings anticipated from integration.

By building this sort of system, AT&T would draw the licensees into a closer "working relationship" with American Bell and "promote uniformity of methods" throughout the industry. Even more important, the establishment of "one great exchange system," as Hall put it, would "strengthen all points [of the business] against the dangers of competition."[6] Mindful of the need to begin preparations for the post-patent era and the reemergence of competition, Hall wanted to invest his enterprise with the same distinct advantages enjoyed by its erstwhile rival, Western Union: an extensive, nationally integrated network linking together all the major centers of economic activity and a service whose value and scope would enable the company to command a premium price in the marketplace. An AT&T network along these lines would almost certainly ensure the Bell interests' undisputed leadership in the field long after the patents had expired.

With these sorts of great ambitions but with little fanfare, AT&T entered its first major undertaking, a line from New York to Philadelphia. The participation of American Bell management in this project actually predated the organization of the new subsidiary. As a first step, connecting local telephone companies were asked to determine the extent to which they planned to get involved in building and financing the line. Directors of the Metropolitan Telephone and Telegraph Company, American Bell's New York licensee, quickly informed Vail of their intention to share fully in the "profitable results of the new business"; but they pressed for a less ambitious and less expensive venture than the one Vail had in mind. As a compromise, AT&T and the Metropolitan Telephone and Telegraph Company agreed to share equally the costs and revenues of an operation that would be built for about $200,000.[7] There were sharp differences over the manner in which the projected line would be used, but despite this simmering controversy, Hall and his staff succeeded in negotiating over two hundred right-of-way contracts with municipal authorities and in launching their project.[8]

The line was built along what became known as the lower route, a winding path that made its way from New York City to Jersey City over submarine cable; stretched to Bergen Point and through Bayonne on poles belonging to the New York and New Jersey Telephone Company; turned east to Port Richmond, Staten Island; jumped back to New Jersey, again using cable; followed the coastline to Perth Amboy; crossed the Raritan River by cable to South Amboy; continued south along the highway to Smithburg; and ran through the towns of Madison, Marlborough, and Manalapan and from there through Prospertown, Jobstown, Mount Holly, Roncocas, Masonville, Morristown, and Camden; until it finally reached the outskirts of Philadelphia. Estimated construction costs along this particular route were about $255,000, some $30,000 less than the next best alternative. The lower route also employed less cable, reducing one of the most troublesome causes of transmission interference.[9]

The completion of AT&T's twenty-five-circuit line in April 1886 was hailed in the press as a "striking" achievement. The *New York Commercial News,* delighted with this unprecedented technological feat, generously predicted that in "time telephonic communications between New York and London, or even San Francisco and Paris [would] be possible."[10] Other published accounts claimed that the new line had placed AT&T's subscribers in Philadelphia in direct contact with over 15,000 stations in New York City, New Jersey, and Brooklyn.[11]

As it turned out, however, these accolades were somewhat premature. In attempting to hook up with its licensees, AT&T encountered the same sorts of problems that had earlier discouraged the development of the less complex, short-haul toll business. The Metropolitan Company reported "considerable delay in completing [the] city connection" upon which access to its subscribers depended because of unanticipated "underground difficulties." In the spring of 1886 its management pointedly told Hall not to expect such connections anytime soon. The situation in Philadelphia was even worse. According to Hall, AT&T's lines had reached the boundaries of that city, but officials of the Pennsylvania company had not shown "any disposition . . . to cooperate" in the provision of "proper connections."[12] Two years later, a considerably more frustrated Hall would complain that "what business we have so far developed has been obtained at great expense of constructing at the Philadelphia [end of] our own terminal system." "Neither the business nor the assistance" once expected from the Philadelphia Company had materialized; in light of these disappointments, "the purpose for which the line was intended [now seemed] practically defeated."[13]

Hall and Vail were both baffled and disappointed. From the outset, a bewildering variety of unforeseen problems had arisen. Officials of the local company in Philadelphia had balked at concluding arrangements to collect toll charges from subscribers claiming that this would automatically make them liable for bad accounts.[14] Western Union, which as a result of the patent settlement of 1879 still held an interest in both the New York and Philadelphia exchanges, had taken upon itself to greatly complicate matters whenever possible; this was one measure of the telegraph company's growing anxiety over the prospects of competition with AT&T in the long-distance message market. In September 1886 a frustrated Hall wrote Vail that he was "very decided[ly] of the opinion that while it may be, and probably will be possible to make a contract for terminal arrangements with the Philadelphia Company, no satisfactory service will ever be obtained so long as that company is managed by the Western Union Telegraph Company."[15]

An even more serious impediment to operational integration, however, was the poor physical state of the local plant. After two more years of little or no progress on this front, Hall finally concluded that

> the grade of service is so low and the condition of the plant so defective in every exchange system with which we connect, that today we cannot give even a fair grade of service between any two exchanges. Our business has been limited to what I have always deemed its least profitable part: leasing of private lines. We are just commencing to get some terminal and branch connections, and no one who has become familiar with the conditions of our business can doubt that if the connecting exchanges were thoroughly equipped (as they should be for their own preservation) with lines and instruments suitable for our service, none of our existing pole lines would be large enough to carry the wires needed for our business.[16]

The bad news could be read in AT&T's balance sheets. Earnings plunged well below expectations as AT&T's management struggled unsuccessfully to connect its lines with the local plants owned and operated by the licensees. On revenues of $146,555 (1888), AT&T reported a net income of only $21,768. The following year, its net return increased, but it was still only 1.75 percent on an investment valued at about $3 million. Operating expenses accounted for a hefty 85 percent of total income in 1888 and just over 80 percent in 1889.[17] AT&T officials were, nevertheless, determined to make the long-distance business a success, and they continued to pour capital into the undertaking. Management even broke with its deeply ingrained conservative tradition by floating $2 million in 7 percent, ten-year debentures to underwrite new lines between Albany and Buffalo, between New York, Boston, and Providence, and between Chicago and Milwaukee.

In part the problem was technical. Local telephone company plants, which at the time largely used single, grounded iron wires, were found to be generally unsuitable for connection with AT&T's higher-grade, two-circuit metallic network. Under the best circumstances, it was a considerable feat to tie these systems together without dramatically compromising the quality of electrical communication on both.[18] Of course, the licensees could have upgraded their plants. But most of them simply did not command the resources necessary. Only the Metropolitan Telephone Company of New York had undertaken anything resembling a major (though not entirely successful) effort to connect AT&T circuits into its own exchange system; at the last minute, the Metropolitan Company had, to this end, made costly modifications in its mammoth six thousand-station Cortlandt Street switchboard.[19] More often than not, however, such endeavors had been deferred, the victims of tight budgets and widespread indifference at the local-company level.[20]

American Bell and AT&T officials witnessed these developments with increasing concern. The establishment and expansion of AT&T's long-distance system was forcing a dramatic transformation of the business and raising important questions regarding the manner in which they had conducted their affairs to date. Operational interdependence compelled company leaders to reconsider the carefully drawn managerial jurisdictions established during the consolidations of the early and mid-1880s. The expense and techni-

The Cortlandt Street switchboard, at the time of its completion the nation's largest, with a capacity of up to ten thousand lines. The switchboard was modified extensively in 1886 in an attempt to accommodate the technical requirements of AT&T's long-distance connections.

cal complexity of interconnection had begun to overwhelm licensee resources. For Bell officials, the big issue was whether these circumstances justified a change in established policy—whether leaving the planning, construction, and administration of the local segment of the network in the hands of the local companies continued to make sense. This question divided Bell's management into two camps as the decade drew to a close. In one camp were those who continued to support the policy of keeping American Bell as simply a holding company without major managerial or financial responsibilities in telephone operations. In the other camp were those who advocated a much broader, more active corporate role for their company. Their ranks harbored the spokesmen for the relatively new concepts of engineering efficiency, those convinced that when the patents expired, the company would have to be able to depend on its effective telephone services and operations in fighting off the competition that would emerge.

Predictably, the new outlook was strongest among those deeply involved in AT&T's affairs, and the earliest expressions of this position came in response to policies that they regarded as a step backwards in the evolution of the company. A proposal that floated around the executive suite in 1886—to turn American Bell into a consolidated holding company—sparked just such a response.[21] The plan's author, Charles Bowditch, was then a director and vice president of American Bell. The opposition was led by E. J. Hall, the long-distance company's leading expert on telephone operations.

In a lengthy and persuasively argued opinion, Hall pointed out that the Bowditch plan was hazardous and troubling in its political ramifications. It contemplated the creation of the nation's most visible monopoly and greatest concentration of corporate wealth. The new holding company, its ranks swollen with local-company personnel, its holdings filled with watered investments from the local companies, and its treasury overloaded with a capitalization of $75 million, would undoubtedly inspire the ire and intervention of public officials, making it a favorite target for every demagogue on the election trail. Existing conditions, Hall said, required Bell officials to consider the public temper as much as "the equities involved" in implementing such a reorganization. Lawmakers would not long tolerate the existence of such a mammoth enterprise before "swiftly crippl[ing] it so that not a dollar could be paid in dividends."[22]

Such a massive realignment of properties and ownership would be troubling in other ways as well. Of particular concern was its impact on Bell licensees, the entrepreneurs whose efforts and capital had been largely responsible for building the business and whose allegiance was needed to sustain its growth. As a group they had suffered through several changes in policy and corporate structure in the recent past, and Hall now believed that it was time to reduce their uncertainty, not increase it. With the capital requirements of telephony increasing at an explosive rate, American Bell could not afford to lose the "good will" and "cooperation" of its licensees; nor could it stand the loss of experienced officials, many of whom would resign rather than take orders from a centralized company operating out of Boston.[23] The firm had cultivated these relationships; in policy and practice it had encouraged the recruitment and development of an entrepreneurial and independent class of managers to run the local end of its business. The licensees were a valuable resource not to be sacrificed lightly.

Still, Hall did not rule out some form of consolidation. Having encountered firsthand a variety of problems in managing AT&T's operations, particularly in the realm of licensee cooperation, he, too, acknowledged the need to strengthen the overall coordination and administration of the enterprise; and he knew that it was necessary as well to diversify its sources of income. But for him, it was AT&T, an operating company, not American Bell, that appeared to be the most promising vehicle for managing such changes. After all, AT&T had no ceiling on its capitalization, giving it the financial muscle to consolidate the business when and wherever necessary. And it already enjoyed a close (if not always satisfactory) working relationship with many connecting licensees.

Hall, however, was not ready to rush to judgment on just how soon or how far such a consolidation should go. The goal of a more centralized control, he argued, was best achieved through a process that proceeded at a gradual pace. Simplest to orchestrate would be a gradual transfer and exchange of AT&T and licensee stock, a move that could be made slowly so as not to arouse a public controversy. Over time, AT&T could acquire a majority position in the licensees and provide local investors with holdings in the new parent enterprise. "In this way a final consolidation could be eventually effected. Every step in the transition would be easy and natural. The change would be too gradual to attract notice."[24]

Hall speculated that during the interim, both American Bell and AT&T could continue to operate as discrete businesses "under one direction," the first "devoting all its forces to the defense of the present patents, the acquisition of new ones, the collection of royalties and the observance of . . . contracts by the licensees," the second organizing "with the strongest practical telephone men available, for active work in the field." While American Bell piled up profits for the shareholders from instrument royalties, AT&T would push the business "in every possible direction," using the local companies to secure favorable legislation, "control public opinion," maintain rates, and "get everything into the best possible condition to face . . . competition." This two-track policy, if successfully implemented, would spare Bell a "serious . . . loss of profits" as it slid through the transition of the 1890s.[25]

Expectations regarding the promise and profitability of the local exchange and short-haul toll business had not panned out. Several licensees, particularly those that had vastly underestimated the requirements of their increasingly capital-intensive operations, were now suffering serious financial stress; many more had requested of American Bell a measure of financial support. The Forbes administration had not counted on these developments. Indeed, from the outset, the company had assumed that in the not too distant future the licensees would become a source of income for the parent enterprise, not the reverse. The pressures of plant expansion and modernization, however, had changed the outcome. And as American Bell and AT&T officials continued to urge licensees to upgrade and accelerate their construction programs, a movement began in the field to reform what many local officials regarded as unfair licensing and royalty practices.

In Gardiner Hubbard the discontented found an articulate spokesman. After poor earnings in some areas of the business forced the New England Company to raise its rents in 1885, leading to a dramatic reduction in subscriptions, Hubbard took up the licensees' case and demanded of Forbes a full accounting of these events and their causes. "The American Bell is not an ordinary manufacturing company, it is to a certain extent a quasi public corporation,"

Hubbard reminded Forbes. "It cannot like an ordinary manufacturing company carry on its business simply with a view to the largest profit to its stockholders, but is bound to consider the rights of the public. If the charges are too high, then the public have a right to complain, and will complain, and will be heard as well through the press as through the legislatures of several states. That charges are too high is shown by the large dividends we declare."[26]

Forbes saw the problem in a different light. For several years American Bell had voluntarily deferred the dividends due on its holdings in the licensees; only recently had the company begun to demand income from this source. Admittedly, in some cases the payments had involved a return of 15–16 percent; but the average payout for the years 1879–85 amounted to barely 8 percent. Moreover, its effect on the cost of telephone service seemed hardly extraordinary, ranging as it did from two to four cents a connection.[27] The specific increases about which Hubbard complained, Forbes said, were designed to correct large losses suffered by exchanges located outside of Boston. In principle, his administration had always opposed subsidizing the New England Company's marginal operations with profits earned on its more successful ventures. The rate increases merely represented what he believed were the necessary steps to redress this imbalance.

Of course, even Forbes acknowledged that such rate increases provoked the ire of unhappy subscribers and would, if sustained over the long run, weaken the competitive position of the local telephone companies. Thus, before he resigned as president of the American Bell Company, Forbes promised Hubbard and the concerned licensees that his administration would look "over the whole field with a view of making changes in the contracts based on some principle which can be applied fairly in various classes of cases."[28] But nothing came of this—in large part, it seems, because a powerful and conservative faction within Bell's leadership circles was more inclined to strengthen the finances of the parent company than to prepare affiliated local interests for competition. This conservatism prevented the company from responding more creatively to the problems slowly developing in the field. In 1887, Forbes ended his term as American Bell's president, and Howard Stockton, who took his place, appeared no more inclined to introduce such changes than had his predecessor.

The financial problems of the licensees did not, however, just dwindle away. Indeed, they became more complicated and more serious as the decade drew on. Hubbard estimated that of the $40.5 million in stock listed on their books, only about 33 percent, or $13.65 million, represented tangible assets such as plant and equipment. Aggregate local-company earnings for 1887 constituted an average return on equity of only 5.5 percent, although the diluted nature of the stock meant that the returns on actual investment were somewhat higher.[29]

There were all too many disparities in the proportion of total income that licensees paid for leased telephone equipment. Rural phone companies typically earmarked as much as 23 percent of their revenues for such payments, urban-centered exchanges rarely more than 16 percent. Ironically, the companies least able to support such expenses paid the highest fees, owing to an unwillingness on the part of American Bell's officials to make "allowances" for differing prospects and market conditions. Hubbard pointed out this problem in a letter to Stockton (see table 1). As a consequence, he concluded, expansion had been stymied and many small exchanges had gone out of business.[30]

In the spring of 1888 Hubbard proposed that the parent company take measures to redress these disparities by discounting licensee rental payments until such fees equaled a commission of no more than 15 percent on their total receipts.[31] Companies benefiting from these reductions would be required to apply the funds to the extension of plant or to reductions in subscriber rates. At Stockton's suggestion, Hubbard brought his plan before American Bell's board of directors, but there it received an icy reception. As one member of the board observed, Bell had "the whip-hand over the telephone business of this country"; concerns for current profits, and not the firm's long-term position, ruled the day.[32]

American Bell officials chose to ignore the mounting financial problems of the licensees. But each extension of AT&T's long-distance lines increased the pressure to resolve the company's financial, organizational, and technological difficulties. On the technical front, there was an obvious need for greater uniformity of practice and standardization of equipment throughout the industry. As Hall pointed out in 1889, the lack of proper terminal facilities at the exchange level was blocking commercial development across the

TABLE 1. THE IMPACT OF AMERICAN BELL ROYALTY AND DIVIDEND POLICIES ON LICENSEE FINANCIAL RESULTS, 1881

Exchange Class[a]	Average Charge per Subscriber	Royalty		Total Earnings	Royalty	
		($)	(%)	(000)	(000)	(%)
1	$85.00	$11.56	13.0%	$5,365	$869	16%
2	45.40	11.68	26.5	2,458	562	23
3	48.25	12.03	25.5	1,102	258	23

Exchange Class[a]	Capital (000)	Dividend		Royalty (000)	Royalty as Percentage of Dividends
		(000)	(%)		
1	$20,791	$1,357	6.5%	$869	65%
2	13,784	639	4.1	562	90
3	5,950	244	4.1	258	108

Source: Data compiled from Hubbard to Stockton, 7 April 1888, AT&T Archives, box 1115.

[a]Exchange class refers to the number of subscribers served. Class 1 exchanges were the largest, serving urban areas; class 3 exchanges were the smallest, serving rural areas.

entire field of long-distance electrical communication. This contrasted sharply with American Bell's status in the patent arena. There its already strong position had actually become more powerful as a result of the acquisition of Western Electric and the provisions written into the license contracts requiring local companies to surrender to American Bell any inventions pertaining to the electrical transmission of voice arising out of their own experiments. But ironically, Bell officers could not force their local affiliates to adopt these innovations. As Hall frequently reminded his colleagues during this period, in an unexpected twist of fate, AT&T was being forced to make up for the deficiencies of the operating companies. He maintained that "shortcomings" of this type were "not due to a lack of willingness to carry out the contracts, but to ignorance of what is needed and to bad methods of organization and management which can be corrected only by education and gradual reorganization. As a rule, local managers believe their service to be good enough for all practical purposes, and it's only by showing them something vastly better and then keeping the pressure on day by day to raise their standards that improvement can be effected."[33]

If Hall had his way, operational integration would be the centerpiece of American Bell's commercial strategy, and standardization, the technical tactic for implementing it. Among the company's technical staff, standardization had already become a constant refrain,[34] but the disappointing results of AT&T's endeavors to connect up with licensee exchanges demonstrated how far the company had to go to achieve that goal. Within the firm the advocates of thoroughgoing integration were gathering strength. At the Eleventh National Exchange Conference, convened in 1889, J. J. Carty delivered a seminal paper entitled the "New Era of Telephony." The paper described an age in which "proper engineering" practices were implemented by a trained and experienced staff following plans and procedures developed by a single, unified management.[35] The new era, it said, would witness the elimination of differences between the methods of the East and those of the West. Problems of uneven development would vanish. The needs of the overall system would triumph over highly particular, local demands.

For the present, however, the new era could not unfold, principally because the crucial organizational and technical innovations required for its development had been postponed. Managerial inertia, the enemy of long-run business vitality, had largely prevented the enterprise from adapting quickly and completely to the industry's fast-changing technology. But American Bell's leaders could only temporarily ignore the changes taking place; they could not prevent transformations in their firm's economic, political, and technical settings from forcing the Bell companies to adjust to their new environment. When they did so, it became apparent that the debates of the 1880s had done a great deal to clarify the needs of the new era and to outline the difficult choices that management would have to make.

CHAPTER 7 ❧

The Hudson Years: Centralization and the New Era of Competition

For American Bell, the 1890s marked the end of an era. In 1893 and 1894 the original Bell telephone patents expired, and with them went the durable foundation upon which the firm's monopoly had been built. A financial panic in 1893 temporarily shielded the company from the pressures of competition. But by the end of the decade, investor confidence revived, rekindling widespread interest in the prospects of commercial telephony. Thereafter, a vigorous independent telephone movement took root; competition became a fact of life.

The new era would demand a new outlook and a new style of management on the part of Bell leaders. Yet these were changes that would come slowly to an enterprise whose policies had been and continued to be dominated by a group of strong-willed entrepreneurs and investors. The top position at the company was now held by John E. Hudson, who at first glance seemed indistinguishable from his Brahmin colleagues. A Harvard-educated attorney later purported to have written some of his corporate correspondence in Greek, Hudson would earn the reputation of being a distant and difficult man.[1] For his company's competitors he displayed little more

John E. Hudson, American Bell Telephone's first general counsel. Hudson replaced Vail as general manager of American Bell in late 1886 and as president of AT&T in September 1887. On 1 April 1889 he became American Bell's third president, a position he held until his death in the fall of 1900. Long viewed as a difficult and temperamental man, Hudson orchestrated American Bell's difficult transition from a patent monopoly to a competitive enterprise. Throughout his tenure in office he was plagued by poor health and an uncooperative Executive Committee.

than disdain. Like his predecessors, he would heartily endorse the advice of his company's patent lawyers, who sought primarily through vigorous litigation to defend American Bell's interests and market position.[2]

But Hudson was, nevertheless, a new type of leader for the enterprise. Unlike Hubbard or Forbes, whose extensive holdings were their primary claim to a position of leadership in the business, Hudson had risen to the company's top job through the ranks of management. He had served a distinguished stint as American Bell's first full-time general attorney and had taken on an assignment as president of AT&T following Vail's resignation in 1887. Hudson's experience in both the legal and operating ends of the business appeared to qualify him uniquely for the task of overseeing his company's difficult transition into a competitive enterprise. For the first time in its history, Bell's chief executive was a full-time professional.

Hudson would adopt a cautious, incremental approach to many of the challenges facing his administration. There was no known blueprint, no model, for the kind of corporate transformation that he was destined to oversee. In the area of earnings, his success would depend on how quickly he was able to replace declining instrument royalties with new revenues generated by AT&T's long-distance operations and dividends from the licensees. In the area of administration, it would depend on how soon and how effectively he was able to introduce modern methods of management and how much more efficiently such new techniques would enable the enterprise to operate as an integrated whole.

Fortunately, Hudson would not be alone in inaugurating these changes. Within the business there was already a corps of experienced officials espousing the credo of modern administration. There was already a group advocating—indeed planning—the transformation of American Bell into a company whose income and profits were derived, not from royalties, but from operations spanning the continental United States. Among those taking the lead in introducing the techniques of modern management throughout the business were men like Edward J. Hall and J. J. Carty. In 1890, before the twelfth meeting of the National Telephone Exchange Association, E. J. Hall delivered a seminal paper describing what he believed to be the universal principles governing the structure and operation of the typical local telephone enterprise. The purpose of his address was to foster among the licensees a better "understanding of the benefits . . . derived from properly constructed corporations" and to "formulate," in general terms, the structural specifications of such organizations. To Hall, the modern, large-scale business institution was a complicated "mechanism," an "artificial person," the "construction" of which depended both upon the "legal specifications" described in its charter of incorporation and upon the character of its business. The firm's "physical," or administrative, structure was designed to divide the overall task—in this case that of providing public telephone service—into specialized subunits of labor each representing a discrete function or group of related activities; its "single central authority" combined these various departments into an efficient, coordinated enterprise. The structure of association between specialized functions and the more general task of the modern organization shaped the precise form the lines of communication and authority took; and according to Hall, such links were defined by the needs of the business, not by personal relationships. To facilitate an understanding of these needs, local officials were encouraged to draw their own lines of authority in schematic detail and compose "written [job] specifications" explaining the "exact functions" and responsibilities of each employee. Specialization of function, Hall emphasized, was the organizing principle of the modern business structure.[3]

Hall provided those at the conference with a model of company administration against which they could compare their own organizations (see fig. 1). Based on the departmental structure established at the Metropolitan Telephone and Telegraph Company of New York, the model delineated in detail the lines of authority and communication extending from the shareholders, through the board of directors, to the president and his general manager and then down to the functionally specialized branches of the business. Of fundamental importance was the separation of general corporate functions from those operating activities directly related to the provision of telephone services. For example, the company auditor, treasurer, secretary (in charge of corporate records and correspondence), and legal counsel reported to the president's office. Those in charge of

Figure 1

TERRITORIAL ORGANIZATION – 1890
METROPOLITAN TELEPHONE AND TELEGRAPH CO.

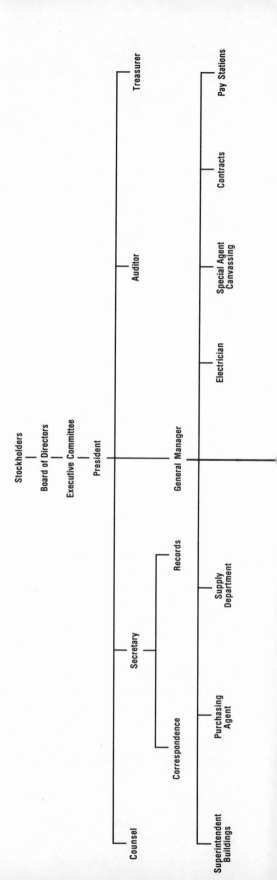

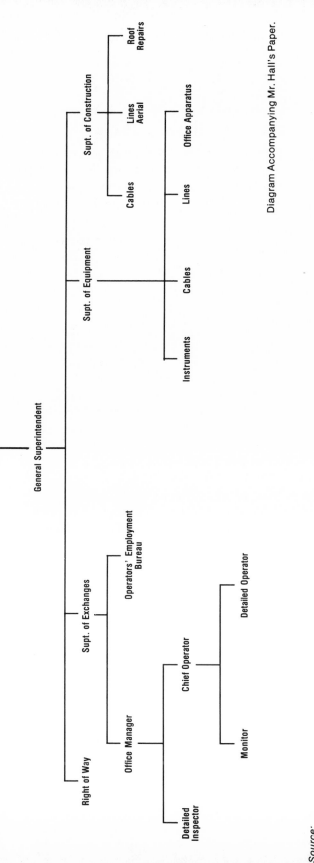

Diagram Accompanying Mr. Hall's Paper.

exchange and toll services were placed under the office of the general manager, which was responsible for building maintenance and planning, purchasing, subscriber canvassing, contracts, pay station collections, and the work of the company electricians. Functions requiring a comparatively large staff were usually organized as separate departments each headed by a specialized superintendent, who, in turn, reported to a general superintendent or assistant general manager. Exchange operators and employees involved in the construction of cables and "aerial" lines; the maintenance of instruments, cables, lines, and "office apparatus"; and the acquisition of right of way were grouped in this fashion. These departments were usually subdivided again along territorial lines to further promote efficiencies of specialization.[4]

Hall's organizational plan parceled responsibility into discrete, identifiable units so as to improve administrative accountability. Whereas the duties of the typical exchange superintendent had once embraced a wide range of activities related to central office operations, the new scheme gave the superintendent control of only one of three major functions: the switchboard operation, in particular the staff of exchange operators. Carty, who had championed the reorganization of Metropolitan's business along these lines, explained to the licensees attending Hall's presentation that the new organization "completely divorced" the superintendent of exchanges from "all responsibility for the plant." Hall elaborated:

> Now, the central thought of specialization is this: that in [every] company one man is responsible for the condition of every line and switchboard in the exchange. There is one man responsible for the physical condition of the property; he has no relation with the operators or with the operations service. That is an entirely separate department, but he is responsible for the operators and subscribers being furnished with suitable apparatus, and apparatus properly taken care of. That simplifies the problem as to who is responsible for any fault in the service. If the fault is one of apparatus or maintenance, then the Superintendent of Equipment is to blame. He has nothing to do with the construction of that plant, and he has nothing to do with the operation of it.[5]

Following Hall's presentation, the participants engaged in a lively discussion over the merits and feasibility of introducing such arrangements elsewhere. Several telephone company representatives questioned the advantages, under existing circumstances, of delegating authority to subordinates and introducing specialization of work assignments. Others noted the difficulties they had encountered in the past in embracing, in a general way, such recommendations. Thomas Lockwood disagreed, maintaining that such problems were frequently surmounted by clarifying internal responsibilities through the sort of organizational arrangements that Hall had proposed. But at best the licensees attending the conference gave Hall's concepts of proper administration only a qualified endorsement. Only the largest of the telephone companies had experimented with the kind of corporate structure advocated by these proponents of modern management. Even Carty's Metropolitan Company, surely among American Bell's most advanced telephone operations, had "not thoroughly completed" adopting the new "scheme" at the time Hall presented his views on the subject.[6] The record was incomplete, the results too preliminary.

Still, among the large, urban-based telephone companies, the ethic of operational specialization gradually took root. Pressed by engineers like Carty—many of whom were inclined to subdivide human activities in much the same way that they delineated the functions of complex machines—licensee officials came to accept the economic and administrative advantages arising out of what became known as the "territorial" form of organization. By 1907, few companies had not adopted some version of the territorial scheme. Only in the smaller exchanges, manned by individuals who for economic reasons were engaged in more than one assignment, was the adoption of these specialized arrangements delayed.

It was during Hudson's term that overall systems performance began to take hold as an important consideration among local-company officials. From the outset this shift in outlook was limited in scope and voluntary in nature. But it marked a development of considerable significance, for it hastened the day when operational integration of long-distance and local exchange services would become a reality.

The new outlook was most evident in the changing technical standards adopted at such interfirm conferences as the National Telephone Exchange Association. In the late 1880s the association had been divided into two specialized subcommittees for dealing with the different problems associated with cable transmission and switchboard systems. By and large, however, these committees continued for some time to focus almost exclusively on matters of local interest, such as operator staff accountability or the economic characteristics of various exchange signaling mechanisms. Problems such as the conversion of central office facilities to metallic-circuit technology, the foundation of local and long-distance service integration, were ignored. But by 1891 there were signs that a new perspective was beginning to emerge. While the efficiency of local transmission was still a subject of concern, the clearly stated objective of the cable committee meeting convened at that time was to provide the best transmission system for general use—that is, for long-distance as well as local service. Trade-offs were accepted in order to arrive at the most efficient overall choice.[7]

The same attitude also came to prevail at the switchboard committee meetings, which had replaced the general exchange conferences as a specialized forum dealing with advances in central office technology. Now the Metropolitan Telephone and Telegraph Company's experience with the problems of electrical imbalance on switchboards during metallic-circuit conversion (to facilitate long-distance service) attracted considerable interest.[8] It would be some time before Bell would have a standard switch that satisfied the transmission requirements of both local and long-distance connections; but affiliated companies seemed to be doing what they could to promote the efficiency of the entire Bell System—not just their local business. Telephone company officials displayed a new willingness to help AT&T use more efficiently a toll-service plant that by this time represented an investment of about $4 million.[9]

Working through conferences such as these, Bell officials gradually established many of the technical standards needed in their industry. These specifications were arrived at by consensus and were frequently adjusted for local conditions. Compliance continued to be of a voluntary nature, but as Hall pointed out, American Bell earnestly hoped that the recommendations of its technical staff and conference attendees would not be violated without sufficient cause. "The whole purpose of this specification," he explained,

> was to furnish a standard basis for cable manufacture: and the point has been made before . . . that it does not absolutely follow that

every manager shall buy cable on this specification for every place, but that having the specification he is on the defensive. If [extraordinary conditions prevail,] it is for the engineer of the [local] company to say to his Board of Directors, or to his superior officers that the specification should be departed from to provide for that special case: but if he makes such a change as that, whoever makes it is on the defensive: he is obliged to prove that sufficient cause exists for varying the standard specification.[10]

Although the pace of technical standardization would prove frustrating to those eager for more rapid progress, on the whole the company had good reason to be pleased with the results registered during the early 1890s. The most notable signs of progress were the figures depicting the "modernization" of local-company plant, in particular those indicating the annual increases in subscriber stations hooked into converted or newly constructed metallic-circuit central office systems. At the end of 1891 only 11,584 out of 216,017 telephones had furnished metallic service. A year later that number had climbed to 23,053.[11] As Hudson explained to the shareholders, the increased use of metallic-circuit systems was of vital importance to the company: "The business or professional man, aside from the fact that he finds this class of service more convenient even for local use, is fast coming to consider the advantage which it affords of placing him, almost on the instant, in communication with the numerous cities and towns through some twenty states of the union, which have successfully been brought within reach of his voice by extension of the Long Distance System."[12]

The process of standardization—the centerpiece of modern, large-scale technical and operational integration—was extended into the realm of administration as well. Here the goal was much the same as in operations: to improve the overall performance of the business by enabling headquarters management to monitor more effectively the affairs of its various business interests. During the 1890s the most notable progress on this front was made in the area of intercorporate accounts.

In the seven years since the accounting revisions of 1884 there had been only marginal advances in the breadth and quality of the financial information that the parent company received from the field. At that time, American Bell's Auditing Department, under the direction of Thomas Sherwin, had conducted an extensive survey of prevailing bookkeeping practices among the licensees and had then introduced the industry's first uniform accounting system.[13] Sherwin's system—which replaced the "ordinary double entry system" designed by the company's first auditor—involved relatively sophisticated ways of handling depreciation charges, toll service accounts, assets, and operational expenses (which, for instance, were divided into sixteen categories instead of the original three). In 1887, minor revisions had been made in the local damage and rebate accounts "to secure greater uniformity" and to facilitate a "comparison of results in different localities." American Bell's purpose was to obtain monthly reports that furnished not only "the record of instruments and leases, but also the inventories of property of different classes." As Sherwin once explained to the licensees, his standardized system of accounts, "while not too complicated, should be capable of showing, at the end of each month, the true condition of the company's affairs, and the actual earnings and expenses pertaining to the several departments of the business."[14]

But that goal had not been achieved. A year after Hudson became American Bell's third president, a departmental examination of the monthly reports submitted by the licensees turned up widespread irregularities in the classification of certain assets and liabilities. Differing local conditions, together with the "lack of a common basis of comparison," accounted for these discrepancies; but instead of suggesting that the Auditing Department once again devise a detailed reporting format for the field, Hall argued that this time Bell should establish "certain general classes into which all accounts can be combined." His major concern was that the books be designed in a manner useful "for study," and he recommended that an inventory of "operating statistics" accompany each submission, since these figures were "of such importance that [management] ought to have them always in mind."[15] The number of telephones installed or removed, the net gain or loss in subscribers, local-company construction expenditures, and the number of metallic circuits added to the local wire plant were among the items that Hall considered to be of special importance to those planning the financial and operating activities for the Bell enterprise.

In 1891 Sherwin's office introduced a new accounting system incorporating many of Hall's recommendations. The three primary objectives were:

First. That the books may be made to contain in themselves a complete and clear record of all financial transactions of a company, in such form that any competent accountant can readily trace each item from its first entry upon the books to its final place in the proper ledger account, or reverse the process; and that the books shall afford the means of verifying easily, and with the utmost certainty attainable, the accuracy both of the accounting and the money departments.

Second. That from the book accounts, statements can be made up without difficulty, showing, at the close of any month, the property of the Company, its liabilities, and the results of the business, either in condensed form or with such analysis and detail as may be desired. With a proper classification, the monthly trial balances will present the accounts in such form that a comparison of successive sheets will afford a reasonably accurate history of the business, month by month, show the increase or decrease of different kinds of property or obligations, and draw attention to unusual amounts in any of the principal classes of earnings or expenses.

Third. That the method should be simple and direct, requiring no unnecessary expenditure of time and labor, adapted to the needs of all the companies operating the various branches of the telephone business, whether employing a large and thoroughly organized force of accountants or otherwise; and that its adoption may be productive of such uniformity in the accounts of the several companies as will afford the best facilities for the comparison of results.[16]

Sherwin's innovations made the accounts a more useful tool for monitoring the financial condition of Bell licensees, and the revisions that he introduced reflected management's new emphasis on operations (as opposed to royalties) as a source of revenue. Right-of-way accounts, separately identified since 1884, were consolidated into a new franchise account, and the old bells, merchandise, and personal property classifications, which included office furniture, tools, horses, and wagons, were regrouped under

Supply Department assets. Changes in the construction account itemized those assets into five discrete categories—exchange aerial, underground conduit, underground cable, equipment, and toll lines—providing a more detailed picture of the character and condition of operating company facilities. Real estate mortgage notes and bonded debt were added to the list of corporate liabilities, and the old reserve account was subdivided into three categories: maintenance, accrued interest, and taxes.

The most crucial accounting innovations were on the expenditure side of the ledger, where costs were now divided into nine major categories: general, operations, maintenance, rental and royalty, private-line, messenger, real estate, interest, and miscellaneous expenses. In contrast to the 1884 statements (which allocated costs between two general expense and dividend headings and then subdivided them into somewhat ambiguous categories), the revised format simplified the classification of expenditures in a manner that made the records more useful for tracking and comparing the performance of local managements. Salaries, for example, were no longer reported on a consolidated basis; instead, they were assigned to specific functions, such as maintenance, operations, or administration. In all, the new uniform system of telephone accounts employed twenty-eight classes of expenditures, twelve more than the system it replaced.

American Bell's Auditing Department worked closely with the licensees while the new system was being introduced. To help the licensees determine the precise classification of costs, Bell's auditors provided manuals which included those financial formulas to be used in "ascertain[ing] the returns upon Company investment at different points."[17] The Auditing Department also visited the several licensees and assisted them directly in an effort to eliminate or at least reduce the disparities in local-company bookkeeping practices. They were not always successful. As late as 1894 the accounts of the Chicago Company were in "a very unsatisfactory condition," and "certain irregularities and faulty methods" were found in the manner in which the Chesapeake and Potomac Company maintained its accounts.[18] On the whole, however, the new system quickly improved the flow of revenue and operating statistics into American Bell's headquarters. Management could now assemble the data it needed to compare the operations of the local companies and, for the first time, make decisions based on reliable, detailed information from the field.

American Bell shareholders also enjoyed better information upon which to base their investment decisions. By the end of 1892 the company's *Annual Report* included specific information on the number of stations using metallic-circuit services, the amount of money expended on the extension of the local exchange and toll business ($35,737,049.14 since 1885), and the daily "general average" number of connections recorded at each station (8.05 per day).[19] The new system of accounting in effect tightened the links between these several parts of the business which heretofore had been loosely coordinated. Relative to other large corporations in the 1890s, American Bell was still an administratively decentralized enterprise, but under Hudson the headquarters staff gradually acquired the tools it needed to manage the business effectively as a single enterprise. These organizational innovations were, of course, of particular importance following the rise of competition.

After the original patents had expired in 1893–94, independent phone companies slowly sprang up to compete with Bell licensees or to provide service where none existed. The huge profits that many of the entrepreneurs financing these new endeavors believed American Bell had earned and the high prices charged by Bell's associated companies attracted the new entrants. From 1893 to 1902, 3,057 independent and 988 "mutual" systems were established, with the largest proportion centered in Illinois, Indiana, Ohio, Kansas, Minnesota, Missouri, Iowa, Texas, and Wisconsin. Competition forced the Bell companies to cut prices and seek new ways to enhance their operating and managerial efficiency. Throughout most of this period the overall profitability of the firm was adversely affected.[20] In several instances, service standards suffered as well, as local companies attempted to reduce their costs.[21]

As had been the case when Bell Telephone and Western Union struggled for control of the field, competition accelerated the rate of growth and greatly increased the capital requirements of the business. Between 1890 and 1895 the value of Bell-related telephone plant and equipment had only risen from $53.5 million to $70.1 million. During the next five years, however, licensee assets climbed to $155.5 million. By the end of the decade local affiliates were spending $24 million annually on construction. The sudden rush to occupy the field in advance of their opposition placed new and often severe financial strains on the local companies, strains that were soon shared by American Bell.[22]

Under Hudson's direction, American Bell resumed the practice of extending short-term loans to its licensees in support of their local projects. Most of these loans were later converted into equity. By 1900 the company's holdings in the local businesses had increased by some $39.3 million, and its total amount represented a little over 42 percent of all outstanding licensee stock.[23] In this manner, American Bell gradually advanced its stake in its associated telephone companies. Consolidation of ownership proceeded apace with the technical and operational integration of the system and with the centralization of its overall administration.

But the price of ownership was high. By 1894 American Bell had virtually exhausted its state-authorized capitalization of $20 million. The company was compelled to return to the Massachusetts legislature, this time to petition for a new ceiling of $50 million.[24] Some of this money was to be earmarked for the early retirement of a $2.0 million debenture maturing in 1898; the rest would cover the capital budget of AT&T and local-company construction plans. The bill passed through the legislature with surprising speed, but without warning, Governor Greenhalge vetoed the measure. After the fact, he explained that he not only doubted the need for such a large increase in capitalization but also wanted the company's securities to be governed by the rules that Massachusetts had set for public-service corporations. As a practical matter, this meant that the price of American Bell stock would be established by the state's commissioner of corporations so as to reflect market, rather than par, value—a precaution designed principally to eliminate the distribution to existing shareholders of what Stehman later termed "quasi-stock dividends" in the form of certificates underpriced at the time of purchase.[25]

American Bell officials argued that regulation was unnecessary, but Greenhalge stubbornly stuck to his original position. Hudson was disturbed of course, but eventually he decided that his firm's financial exigencies outweighed his political reservations.

While keeping open his option to move American Bell to another, more hospitable state should the "experiment" in regulated financing prove unsatisfactory, Hudson reluctantly acquiesced to the governor's terms. The legislature promptly passed a new bill which gave to Bell the capitalization that it desired and to Greenhalge the regulations that he demanded. The governor signed the bill in mid-1894. In November, American Bell offered to existing stockholders the first block of 5,000 new shares, of which only 1,632 were purchased at the $190 price set by the state commissioner of corporations. The remaining certificates were, however, sold on the open market, despite the depressed state of the national economy during these years.[26]

In the next few years the pace of American Bell's financial activities escalated rapidly. A second offering of 10,000 shares was announced for June 1895, and a third block of 21,500 certificates was brought into the market less than a year later. The company's most successful issue came in 1896. The economy was recovering, and stockholders snapped up 11,366 shares at a price of $214 each. By 1898 Bell's balance sheet indicated that the book value of its outstanding stock totaled $25.9 million, a $5.9 million increase, for which over $12.0 million in cash had been received.[27]

Near the end of the decade, the Hudson administration found that it needed even more capital than originally anticipated. For the first time in many years, Bell officials—normally a very conservative group—began to consider debt financing as a means of raising money. By 1898 a healthy rebound in corporate earnings had lowered the cost of borrowing to an attractive 4 percent (a rate that under certain circumstances would be significantly below that paid on corporate equity).[28] In July 1898 American Bell took advantage of this favorable situation by selling $10 million worth of 4 percent, ten-year debentures through two venerable Boston brokerage houses. This successful excursion launched a new phase in the company's financial history, and during the next decade Bell management, pressed by competition, borrowed liberally as the enterprise rapidly expanded its telephone network.[29]

Even so, during the last years of the 1890s, management began to recognize that the parent company would soon have to forsake the troubled political climate of Massachusetts for the more agreeable, laissez-faire environment of New York. It was a move similar to those made by other large companies as their corporate leaders sought to escape state jurisdictions that they thought unfairly shackled their venture's entrepreneurial spirit. For American Bell's officials this decision was undoubtedly a difficult one. Boston had been the cradle and home of telephony for nearly a quarter-century, and the company would be severing its ties to a rich and remarkable past. Yet the future beckoned with an urgency that could not be ignored much longer. If Boston represented the invention, the patents, and the Yankee entrepreneurs, New York symbolized the marriage of big business and big finance. Wall Street was the center of a new national, even international, economy based on large-scale production and modern forms of transportation and communication.

Planning for the move to New York began in 1896 with an evaluation of the financial transactions that such a change of location and identity would entail. As had been the case in the past, political considerations loomed prominently in these deliberations. Indeed, management's decision to place the New York–based AT&T Company at the head of the new enterprise reflected not only the comparative convenience of adopting such an arrangement but also a conscious effort to change dramatically the unfavorable public image that the company had acquired as a large, combative, patent monopoly. Long an advocate of this kind of reorganization, Hall earnestly believed that the combination of American Bell's tradi- tional lines of business with AT&T's long-distance operations would dispose "of whatever adverse sentiment . . . exist[ed] against the Company, by reason of its being a corporation deriving its revenue, in large part, from royalties and paying large dividends upon its capital; [and] having assets consisting in great measure of [franchise] stocks which are understood not to represent actual cash purchases."[30] While Bell officials continued to assert that the public's "widespread and unreasoning hostility to combinations" was unjustified, at the same time they were compelled to acknowledge that such opinions represented a major obstacle to what they regarded as the "natural" course of their firm's growth and development.[31]

Such intangible considerations also influenced the nature of the corporate and administrative links that AT&T—the new parent of the reorganized enterprise—would develop with the Bell li-

censees. The "intimate" relationship established "between the subscriber . . . and the system" was seen as a potentially volatile arrangement, suffused in the emotional sentiments of localism and
susceptible to the undisciplined politics of regulation. Under such
circumstances, Hall continued to believe that the company's traditional decentralized structure enjoyed "some advantage" over a
more centralized approach. It was better for AT&T and the associated telephone companies to avoid projecting the image of a large
"foreign" corporation upon a public increasingly hostile to such
combinations. The "disposition of state and city legislative bodies to
concern themselves with the rates and methods . . . of corporations
furnishing public service" meant that the company's earnings were
becoming ever more vulnerable to the unpredictable swings in public
opinion.[32] The political realities of the moment would, in this
instance, outweigh the administrative and financial demands of the
integrated network.

Amid a flurry of complex financial transactions, the reorganization of the business took place on 31 December 1899.[33] The timing
of this event was not without some significance. That Bell officers
had chosen this particular date to transform their patent enterprise
into an operating concern was as much a symbolic gesture to the
promise and challenge of the new century as it was an ac-
knowledgment that the world within which they had so successfully
operated as a patent monopoly had finally passed into history.
Although American Bell continued to exist until 1921, it was a mere
shell of its former self. Stockholders received two AT&T shares for
every one of American Bell's that they turned over; and the new
AT&T entered the new century with assets of over $107 million.

If the new century brought Bell management relief from the
confining political economy of Massachusetts, there were no assurances that AT&T's adjustment to this new era would be trouble-
free. Shortly after the move to New York a crisis erupted within the
firm's once-stable leadership. John Hudson, whose poor health had
forced him to lighten his responsibilities throughout the 1890s, died
suddenly in office in the fall of 1900. Alexander Cochrane, an aging
member of the inner circle of Boston investors who had dominated
the firm's management since the 1880s, agreed to take Hudson's
place—but only temporarily. Forbes and Hubbard had died two years

before, and the company, it seems, had failed to groom a suitable successor. Left with a conspicuous vacuum at the top, the enterprise nevertheless took its first tentative steps toward a new set of policies that would deal more effectively with the challenges of competition.

Hudson had for a time largely succeeded in minimizing the "shock" of American Bell's transition into a competitive business dependent on operational rather than patent royalty earnings. After a precipitous decline in net income in 1893–94, he had managed to compensate partially for the secular erosion in rental payments with a sharp advance in dividends earned on licensee stock. Throughout this period, however, corporate profits remained below the level reached during the last years of American Bell's patent monopoly, and it was 1897 before they broke the $4 million mark. From 1894 to 1896 the return on total capital had averaged between 6.1 percent and 6.7 percent.[34]

The proliferation of competition was exerting substantial pressure on company earnings by the end of the decade. Unit profits declined after 1894: net earnings per station tumbled from a high of about $34 in 1893 to just under $20 by 1900 (see fig. 2). Although the company's subscriber base expanded rapidly during this period, its marginal return on investment fell dramatically at the very moment the capital requirements of the business were advancing at an alarming rate. Partly counterbalancing these ominous trends were the reductions in per station costs (from $58 in 1895 to around $44 a half decade later) that Bell managers were able to effect.[35] But as the new century began, it was questionable what further cuts in this area could be made without adversely affecting the quality of telephone service. AT&T leaders faced the difficult task of formulating a new strategy for coping with dwindling profit margins without unduly sacrificing the price competitiveness and quality of their offerings. For this they would need decisive leadership. And for such leadership they would look outside the firm.

Figure 2

PER STATION STATISTICS

Bell Telephone System

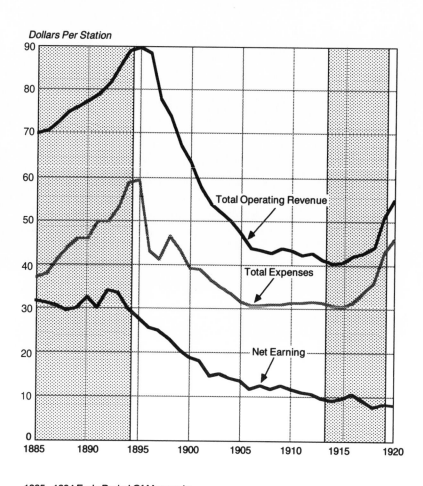

Dollars Per Station

1885 - 1894 Early Period Of Monopoly.

1894 - 1913 Period Of Competition.

1913 - 1920 Period Of "The Kingsbury Commitment"

Source: FCC, Telephone Investigation, p.135

CHAPTER 8 🐦

Interregnum

AT THE TURN OF THE CENTURY the Bell System was in a curious state of organizational transition. Corporate headquarters had been transferred from Boston to New York, but the firm's largest faction of stockholders remained in Massachusetts. Under Hudson, AT&T had acquired large holdings in many of the Bell telephone licensees, but the overall enterprise actually continued to operate in a highly decentralized fashion. Facing increased competition throughout the country, the firm had yet to establish a consistent, long-range strategy for handling such dramatic changes in its marketplace. Hudson's sudden death in 1900 and the appointment of Alexander Cochrane as his temporary replacement did little to ease this situation. AT&T's management continued to deal with competition and with what were now the early stages of a movement in favor of regulatory reform in a piecemeal fashion.

The man chosen to preside as the chief executive of the business as it entered this new and ambiguous world of competition and regulation was Frederick P. Fish, who headed AT&T from 1901 until 1907. His name was probably unfamiliar to many of Bell's shareholders when he took his post because Fish was the first of the company's presidents to be appointed from outside. A patent lawyer with a lucrative practice in Boston and New York, he was certainly well known to the company's Executive Committee. He was a New Englander by birth and, of course, a graduate of Harvard. He had gone on to receive his law degree from the same institution.[1] In practice, his partners had included Colonel Thomas L. Livermore,

Frederick Perry Fish, AT&T's fourth president, the first brought in from outside the firm. At the time of his appointment in July 1901, Fish was considered to be among the nation's leading patent attorneys. During his six-year term in office, he was responsible for introducing important changes in the firm's sub-licensing policies and for supporting crucial changes in the way AT&T's technical staff was organized. Fish resigned in April 1907 after AT&T suffered a sudden and unexpected financial crisis.

1900

The Bell seal, in 1900 used to distinguish not only AT&T's long-distance service but also the entire, integrated operations of the Bell System.

J. J. Carty, chief engineer for AT&T from 1907 until June 1919. Carty began his career in telephony working for the Metropolitan Telephone and Telegraph Company of New York. There he gained an appreciation for the efficiencies of proper organization and became a major advocate of the functional form of administration introduced throughout the Bell System between the years 1907 and 1909. In 1919 Carty became a vice president at AT&T, a position he held until his retirement in mid-1930.

111

former U.S. senator Bainbridge Woodleigh of New Hampshire, and J. J. Storrow, who had served as National Bell's counsel during the company's early patent cases. Fish had long enjoyed a reputation as one of the nation's leading patent and trademark attorneys. At the time of his death in 1930, it was said that he had represented "one side or the other in almost all the great patent cases arising in America" during the forty years of his distinguished career.[2]

His friends at the bar knew him as a persuasive, imaginative advocate with a "wide knowledge of the arts, sciences and history"; a lover of Wagner's music, he sneeringly dismissed modernism—including jazz, *vers libre,* and cubism—as a "reversion to the form of savage tribes."[3] In temperament as well as in outlook, Fish was a member ex officio of that elite society that brought him to the helm of AT&T.

If there was one major difference between Fish and his Brahmin colleagues, it was that his experience and business connections were national rather than regional. He had spent a good portion of the 1890s working as general counsel first to the Thomas-Houston Electric Company and, after 1893, to its successor, the General Electric Company.[4] There he doubtless rubbed elbows with J. P. Morgan and other Wall Street financiers involved in consolidating the electrical industry. His links to Morgan may have helped ease his way into the appointment at AT&T, because the New York investment banker was at that time acquiring a stake in the Bell System. When Fish left AT&T five years later, Morgan and his associates held a majority position in the enterprise.[5] In the interim, however, it was Fish's task to guide a business beset by the challenges of the new era of expanding competition and systems integration.

Internally, Fish's policy was to support the same type of incremental centralization and standardization that had taken place under the Hudson administration. His job in this regard was, however, somewhat less difficult than was his predecessor's. This process of change had developed its own leaders and organizational momentum.[6] Certainly this was true where corporate accounts were concerned. It was likely that AT&T's New York financial connections demanded a better system for establishing accountability on the part of local Bell managers (who were now in charge of greatly expanded investments); but the major thrust of the revisions introduced at this time seemed directly related to competitive pressures that predated the move to New York. Under the existing system, which had been designed during the period of patent monopoly, toll and exchange expenses were bundled together. But as competition developed, it became more and more important to associate earnings and losses with discrete categories of service, to clarify maintenance expenditures, and to develop standard methods for providing for plant depreciation; only then could Bell formulate meaningful pricing strategies and evaluate the comparative return on various parts of the business.[7] The narrower margins stemming from price competition made this all the more important.

Thomas Sherwin's approach to the task of revising the company's accounting system was characteristically methodical and participatory. After holding lengthy discussions with several officials at Bell's headquarters, in January 1903 he distributed a preliminary proposal incorporating many of their suggestions among the licensees for comment. Local-company officers were asked to review the package with their own auditors and suggest any changes that would "adapt the system to the needs of your company and . . . accomplish the objects for which a general system of accounts should be designed."[8] Sherwin also called in the respected public accounting firm of Haskins and Sells for advice on the contemplated revisions.

Over the next twelve months it became apparent that it would be next to impossible to satisfy the licensees on matters of expense and implementation, to meet the detailed reporting requirements of the accounting profession, and to provide AT&T's management with all of the specific information it needed. Haskins and Sells bemoaned the historic "lack of uniformity in the accounting practices and customs of the different offices of the various Bell telephone companies," a flaw that had prevented the company from making an "accurate, reliable or intelligent comparison" of the licensees' balance sheets. In contrast to Sherwin, the accounting firm advocated a "more modern and efficient" scheme that would consolidate telephone property maintenance expenses along the same lines as those used in the railroad industry.[9]

The licensees, on the other hand, rejected the approach

proposed by Haskins and Sells as too complicated, too costly, and too radical a departure from their traditional practices. Echoing the opinion of several of his colleagues, E. B. Field, of the Colorado Telephone Company, wrote:

> It seems to me that if we follow out the ideas outlined by you early in this year . . . we will be able to work out a scheme of accounting that is along the line with which we are already familiar, and which plan is much simpler than the one proposed by Haskins and Sells. I do, however, want to put myself on record as feeling that any scheme of accounting that does not carry out the separating of toll business expenditures, etc. from that of the exchange, is defective. . . . I say this for a great many reasons. First I do not believe that the executive officer of any of our licensees can have a proper grip upon the vital and necessary points of the business, unless this separation is made.[10]

But not all of the licensees were even convinced that the advantages of separating exchange and toll business expenses justified the costs involved in implementing a new and more detailed system of accounting. While agreeing that the "basic principle of the plan [was] ideal," A. R. Schellenberger, president of the Pennsylvania Telephone Company, nevertheless concluded that Sherwin's proposal could not be effected "without excessive elaboration of our field organization."

> We believe that in a large company operating a concentrated territory where special work, and largely that special work only, is done by special men and special sets of men, the plan, if followed out with judgment and care, can become feasible and can be made to show approximately accurate results. It is, however, our unqualified judgment that in a scattered territory like ours, operated by combination men and combination floating gangs—where each practical man and each such gang of men are continuously doing all sorts of work and odd jobs falling under the entire schedule of expense and construction subdivisions every day of the month, the best results we could hope to obtain from an attempted operation of the plan would be so far from exact that it should not be put into operation.[11]

Paradoxically, other local-company officials cited the pressures of competition (which had made the innovations necessary) as the primary reason for not changing their accounts. In areas where Bell licensees said that they had been "drilling" themselves "into the strictest economies" in order to reduce subscriber rates, the new accounting system was viewed as too expensive; it would require, managers said, an expensive, unnecessary expansion in the staff.[12]

Despite the strong misgivings expressed in the field, Sherwin was convinced that AT&T needed a more uniform, detailed accounting system. In a letter to Fish dated 11 September 1903, he spelled out, once again, the compelling reasons for moving ahead in this area. The telephone system had "doubtless become one of greater complexity than that of any other public service system of which we have knowledge." In a single company it was not unusual to discover fifty classes of exchange services, "ranging from the highest grade [of] business station in the large city to that of the residence on a ten-party line," including unlimited and measured services, a combination of both, pay station services, private-branch exchange services, and toll, hotel, and rural services. As a result, the licensees were quoting some forty different, graduated rates which frequently had little or no relationship to the cost of services. The expenses in a central office, for example, ranged "from thirty or forty dollars a drop to one-tenth of that amount," in part due to the "infinite variety" of equipment used and the diverse methods of accounting for it.[13] "Doubtless we are all aiming at uniform standards of construction and operating for the same thing," Sherwin contended,

> but it is obvious that to provide for the service of city, town and rural districts in the various sections of the county, and adapt it to the requirements of the several classes of subscribers and the rates they will pay, must involve many standards, at least in respect to the character and cost of construction and equipment.
>
> Our problem has been, therefore, to devise a system of accounts adapted to the use of these forty companies, our licensees, operating under widely diverse conditions, many of them also operating exchanges and lines of distinctly different classes within their own territorial districts.[14]

Sherwin did not expect licensees to follow "with exactitude" the plan he described. But he assured Fish that even if it were only partially accepted in the field, his new system would enable headquarters to detect the basic trends in the business and ascertain the financial conditions of the several companies with reasonable prompt-ness. The existing system, for instance, furnished only part of the information on investments that management needed to determine "what sort of return might fairly be counted on." The new system would improve that situation and provide management in AT&T and in the local companies with the data they needed for a "proper comprehension" of the business.[15]

On 1 January 1904, Sherwin finally issued the long-awaited accounting regulations to the licensees.[16] The new regulations involved a number of striking procedural changes, but the long process of consultation had forced Sherwin to compromise. They were significantly less experimental than they had been in their original form. Similarly, Sherwin had rejected a recommendation by Haskins and Sells that the licensees delegate more responsibility for account-ing to their local-exchange managers (with whom AT&T could deal directly). Instead, he had decided to leave complete control with the accounting departments of the licensees.[17]

As this episode suggests, the process of administrative cen-tralization within the Bell enterprise took place at a methodical, almost stately, pace. Consultation and not command was the tech-nique used for introducing changes in the system. Sherwin had devoted considerable energy to the task of creating a consensus—to working with the licensees and their specialized accounting depart-ments. The improvements in accounting eventually adopted were the results of compromise and persuasion. The licensees still domi-nated the flow of information from the field to the headquarters, but Sherwin could at least take heart that AT&T had successfully changed the form and content of those reports, making them a more useful tool for management.

Similar changes took place in the realm of financial planning during the first years of the twentieth century. In 1901, AT&T compiled its first "provisional estimate" of the total capital re-quirements of the business for the coming year. This basic exercise in planning grew out of AT&T's new role as the principal source of financial support for the licensees. Theodore Vail, who had returned to the business as a member of AT&T's board of directors, recom-mended casting such estimates in the form of a five-year forecast, but management chose instead to concentrate on the more predictable needs of the immediate term.[18]

Even the one-year plan indicated how hard-pressed AT&T would be to furnish the capital the enterprise needed. The company's general manager launched this exercise in planning by first assembling informal construction figures, including projections for the total investment in plant, equipment, and buildings, as well as the amount of "floating indebtedness" carried over and charged against future earnings for each of the licensees and for AT&T Long Lines for the year 1901. The various Bell companies also supplied detailed plans for covering a portion of their own capital expenditures, specifying those parts that they expected to finance internally out of the next year's revenues and those parts that would be financed by increases in equity or in the local companies' debt obligations. This first exercise in financial planning took about five months from start to completion, and the final figures were presented to the company's Executive Committee in May 1901 (following a "judicious" three-month review with the licensees). The results indicated that AT&T itself would have to raise a considerable pool of capital in the year ahead. Out of the $26.7 million budgeted for plant expansion in 1901–2, AT&T's share amounted to $16.7 million, more than half.[19]

As large as these estimates were, they at least gave AT&T officers a better idea than they had had in the past of what to expect regarding their firm's near- and medium-term financial requirements. In helping its local affiliates underwrite an expansion in plant to meet the competition, AT&T would have to raise money in record amounts. Vail, taking a long-term view of these requirements, concluded that the business needed an infusion of at least $200 million in new capital over the next five years.[20] Yet it seemed unlikely that such a large pool of funds could be raised—even under favorable conditions—through stock issues alone. Debt financing, on the other hand, appeared to offer a reasonable alternative. Although his predecessors had deliberately avoided going into debt, Fish decided that the company could no longer afford to adhere to such a conservative fiscal policy. Apparently more concerned about losing further ground in the marketplace than about losing control of the business to bankers, he turned to the financial houses of Wall Street for the resources AT&T needed to fight off competition. Under Fish the company floated a $13 million block of bonds in 1902, using an underwriting syndicate that included the

Wall Street firm of Kidder, Peabody and Company, as well as one of Bell's traditional brokers, T. Jefferson Coolidge, of Old Colony Trust.[21] Two years later, $20 million worth of three-year 5 percent gold coupon notes were sold through Speyer and Company and Lee Higginson and Company, two Boston-based outlets. In 1905 Fish again broke with the company's standard practices by turning to the international market to underwrite the first $25 million block of what would eventually amount to a $50 million issue of 4 percent collateral trust bonds due in 1929. He used Baring Brothers, Ltd, of England, to spearhead this effort. Finally, a blockbuster $150 million issue of thirty-year 4 percent convertible bonds, part of it brought to the market in 1906, capped the Fish administration's decisive move into debt financing.[22]

AT&T's balance sheet was transformed dramatically by these aggressive financial maneuvers. At the time of Fish's appointment, the company's long-term debt had totaled a very modest $15 million, but the ledger at the end of 1906 listed convertible bonds, collateral trust bonds, and short-term note liabilities of approximately $128 million.[23] In the short span of five years Fish had enlarged AT&T's debt by more than a factor of eight. Encountering extraordinary pressures to expand the business, Fish had embarked on a new and relatively risky course. The high fixed costs that the bonds entailed would have to be paid regardless of whether the company's business was good; as a result, AT&T was more vulnerable than it had been in the past to fluctuations in the national economy. Moreover, if additional working capital were needed quickly, AT&T would probably find it more difficult to borrow cash funds on short notice.

Until this sort of problem arose, however, Fish was able to push ahead with the expansion of the system, a strategy he sustained as well by selling more stock. At the beginning of 1902 the total value of outstanding AT&T stock amounted to $104.7 million; by 31 December 1906 it had climbed to $158.7 million.[24] A considerable proportion of the funds derived from these increases in equity (and in debt) went to the licensees in the form of short-term, convertible notes which were used to replenish their dwindling reserves. AT&T's position of ownership in the licensees thus expanded, climbing to a level of about $406.1 million—a five-year increase of approximately $248.6 million—by mid-1907.[25] Under

TABLE 2. EXPANSION OF TELEPHONE PLANT WITHIN THE BELL SYSTEM, 1898–1906

Year	Total Bell System Telephone Plant (in Millions of Dollars)	Increase	
		In Millions of Dollars	In Percent
1898	$118.1	$13.6	13.051%
1899	145.5	27.4	23.185
1900	180.7	35.2	14.183
1901	211.8	31.0	17.200
1902	250.0	38.2	18.053
1903	284.6	34.6	13.815
1904	316.5	32.0	11.229
1905	368.0	51.5	16.285
1906	450.0	82.0	22.278

Source: FCC, *Telephone Investigation,* 138.

Note: For the years 1898–1902 the annual increases averaged 15.48 percent; for 1902–6, the average was 16.33 percent.

Fish's direction, annual investment in telephone plant advanced at a higher rate between 1902 and 1906 than it had during the previous half-decade (see table 2). By the close of 1906, the Bell System's total plant was valued at $450 million—more than twice as much as in 1901. The system was both larger and more tightly controlled by the parent company as a result of these changes.

Less newsworthy than the firm's changing financial strategies but equally important to the process of managerial centralization were the new initiatives undertaken with regard to the system's technological integration. As before, the extension of AT&T's long-distance network intensified pressures for the standardization of plant and operations throughout the business. There was—in the words of one Long Lines official—a "great need" for a central engineering department which would "consider and pass upon the plans of modern telephone exchanges."[26] Neither the Cable Committee nor the Switchboard Committee had succeeded in promoting uniformity in the construction and operation of local facilities. Uneven technological development continued to be a major barrier to plans for a complete integration of exchange and long-distance services. In dealing with this challenge, American Bell's Mechanical Department had been frustrated by its limited authority over licensee operations. More progress had been made following the appointment of Joseph P. Davis as head of American Bell's newly established Engineers Department; but even his organization was empowered to intervene in local-company affairs only after receiving a formal request from the field. Decisions to adopt the recommended standards continued to be made on a voluntary basis by local officials. Western Electric still designed apparatus to whatever specifications the licensees furnished.

What, then, had the parent company's Engineers Department achieved? For one thing, it had collected a vast library of information on actual practices of the licensees, a body of data that would become essential in the formulation of universal standards. Moreover, Davis's annual departmental reports suggested that the licensees, particularly in the more developed eastern sections of the country, were turning increasingly to AT&T engineers for technical advice and exchange facility specifications. But the limits to voluntary cooperation were clear. Indeed, the company's longstanding effort to establish universal standards was in danger of becoming a victim of competition.

Competition had driven down the price of telephone service and in many instances had compelled the licensees to abandon the construction and operations standards advocated by Bell engineers in order to cut costs. Nowhere was this dilemma more evident than in the Midwest. There the local companies had built a frustrating array of five- and ten-party lines, despite persistent warnings from the home office against adopting such inexpensive systems. Later J. J. Carty would blame the additional delay and expense involved in the installation of new equipment on the legacy of such practices. In a letter to H. B. Thayer, at the time the head of AT&T's manufacturing operations, Carty noted that "it had been said, that it cost Western Electric Company about twice as much for the labor involved in making a given switchboard installation in connection with the Chicago Telephone Company plant as it did elsewhere. This was due to the fact that their apparatus had so many special requirements and needed so many special adjustments that special individual attention had to be given to the particular pieces of apparatus which, in other installations, could all be treated alike in classes."[27] The letter criticized local policies that allowed what was described as a "most radical departure from proper standards" and commented on "the incompatibility of many types of apparatus used" in the present plant.

In 1902, in an effort to prevent this sort of problem from becoming more widespread, Fish urged Joseph Davis to take direct responsibility for the development of telephone facilities in the territories served by the Michigan, Wisconsin, and Northwestern Bell licensees.[28] AT&T had obtained a controlling interest in these companies after the financial syndicate that had managed them since the mid-1880s had failed. The new arrangement gave Fish and Davis an opportunity to intervene directly into their disorderly affairs. But the two men disagreed as to how the uniformity that they both desired should be achieved.

Davis sought the consolidation of the several local engineering departments into one large organization. Fish, on the other hand, was reluctant to sanction such a dramatic change in the structure of local engineering functions. As a compromise, they experimented with a new type of organization using engineers from AT&T's central staff but assigning them to positions in each of the licensees' operations. Companies joining this cooperative endeavor agreed to submit formal construction estimates and large supply orders to Davis's local representatives for approval. Certification implied that the projects had been designed and implemented in accordance with specifications established by AT&T's central engineering staff or by the three-member committee of specialists that Davis organized to supervise activities in the field. Although the headquarters group was available for consultation and assistance, Fish assured participating licensees that the main purpose of the new policy was "so far as possible, to clothe the local Engineer with authority to deal with ordinary engineering problems, without referring them to" AT&T headquarters.[29]

Direct responsibility for facility inspection and data gathering was assumed by the new engineering chiefs. The collection of information depicting the local patterns of telephone traffic, the design of central office and local wire facilities, and the condition and operation of aerial and underground projects gave AT&T the data it needed to assemble an account of the business's growth. Statistics on plant utilization and on the quality of service were also valuable in framing the parent organization's plans.

The new arrangements did not entirely resolve the conflicts that had arisen over Bell's engineering authority in the field. Responsibility for the design and maintenance of exchange and short-haul toll facilities (which in the past had rested with the licensees' plant superintendents) had now been given to the engineers appointed by AT&T. But their ultimate source of authority was still a matter of some misunderstanding. Since the cost and kind of plant installed affected the general posture of the associated companies, licensee officials insisted on retaining "absolute control" over decisions relating to the development of their facilities.[30] Fish accorded them a sympathetic hearing, agreeing in principle that the dual responsibility held by Bell's engineers created a somewhat awkward arrangement. Both he and Davis assured the licensees that AT&T did not contemplate a further erosion of local prerogatives. AT&T officials repeatedly emphasized that "neither in theory nor in fact" did the Engineers Department want to exercise broad control over the licensees. The local engineer, Fish explained, was "entirely subject to the orders of the executives of his [local] company"; nevertheless, "it was part of the scheme that those [engineers were] not to approve an estimate unless it compli[ed] with the standard requirements of [AT&T] Engineering." The limits of local discretion were thus carefully framed. Except in cases where a "very serious difference in views" arose between local-company management and the headquarter's technical staff, Fish was inclined to depend on voluntary cooperation and compromise to effect the desired changes. But he was ready to intervene personally should major disagreements arise "simply because of the very large interest" that his company held in the licensees.[31]

As it turned out, about half of the thirty-three local companies participated in Davis's "experiment" in centralized engineering—a number smaller than anticipated but larger than the department was actually equipped to handle efficiently.[32] As a result, in 1904 the task of designing telephone apparatus—the subject of much "inconvenience and delay"—was transferred to the Western Electric Manufacturing Company. This kept the job under one corporate entity; and a member of AT&T's Engineers Department served with two Western Electric officials on a committee that established priorities for and guided the development work.[33] The following year, Western Electric also assumed responsibility for the inspection and certification of telephone construction material, a role related to its expanding activities as the general purchaser, warehouser, and distributor of supplies for the operating companies.[34]

By this time, AT&T's engineering staff had begun to have a profound impact on telephone development. Standardization was becoming the order of the day; the engineering ethic that spread gradually through the enterprise fostered a new, long-term perspective on the costs of conducting the business. First costs, the charges associated with the purchase and initial installation of telephone equipment, no longer constituted the only relevant factor in the equation that managers employed to determine the comparative advantages of different operating systems. Licensee officials began scrutinizing the long-run expenses of maintaining and enhancing the capacity of their operations; they began to bolster their general development plans and capital budgets with fifteen-year forecasts (prepared by the general staff) of population growth, calling rates, and subscriber line requirements. The complex process of choosing equipment began gradually to be systematized and distilled into distinct stages. The technical staff's bias in favor of deliberate analysis and measurement of results began to spread throughout the entire organization.

These changes took place gradually, and at this time only a handful of the Bell System's leaders seemed prepared to move more decisively toward a completely centralized system. Even Davis—the staunchest advocate of standardization and the general engineering ethic—thought that the licensees' general managers were still the officials most familiar with local conditions and best suited to make decisions about the construction and modernization of the exchange and toll plants.[35] Others within AT&T were dissatisfied with these partial results. Hammond Hayes, who succeeded Davis, wrote:

> At the present time our relations with the operating companies are dependent only upon personal good will and the influence and prestige that comes from men well equipped and doing good work. I feel that we have established satisfactory personal relations with the officers and engineers of the local companies and that without exception all engineers avail themselves of such advice and aid as they require and we can give. On the other hand, many of the engineers disregard recommendations and specifications which we consider proper and substitute others on the same subject, many of which are improper and do not operate to the best interests of their own company nor of the business at large.[36]

Hayes had good reason to be concerned about the overall performance of the business. Operational integration had taken place at a pace too slow for comfort. The company's position in urban markets, once seen as virtually unassailable, now seemed threatened by a determined and vigorous competition. Indeed, by 1906 the independents could boast almost as many subscribers as the number claimed by the Bell interests. Clearly, the opposition was spreading at a more rapid rate than originally anticipated. AT&T's inclination to bank its future on the willingness of subscribers to pay a premium for its high-quality, nationally integrated services—rather than to compete strictly on the basis of price—had added considerable momentum to this trend. The willingness of independents to serve marginal, rural markets long ignored by the Bell interests also had enabled them to expand at an alarming rate.

The Fish administration had undertaken a number of initiatives designed to improve the system's overall competitive position. In 1901 the Western Electric Company and the Bell Telephone Company of Philadelphia had begun an experiment in centralized purchasing. Under the agreement, AT&T's manufacturing arm purchased, stored, and delivered to the Pennsylvania company equipment produced and sold by independent suppliers. Though at first glance the savings engendered by such an arrangement were of modest proportions, by 1904 the overall results were satisfactory enough to encourage twelve additional Bell companies to negotiate similar contracts. In 1908 these arrangements were institutionalized on a companywide basis with the introduction of Western Electric's "Standard Supply Contract."[37]

Potentially even more significant was the revision in the license contract fee formula that AT&T had inaugurated in 1902. At that time, the schedule for such charges had been changed from one calculated to reflect the number of instruments in the field to one based on a proportion (4.5 percent) of gross operating company revenues. In offering to make such a modification, Fish had alleged that the new arrangement "involve[d] a substantial reduction in the amounts [they] paid" to AT&T.[38] (The actual savings are difficult to gauge owing to the increases in the number of telephones installed after the new arrangements went into effect.)[39] In encouraging the licensees to adopt these changes, Fish had also emphasized the advantages of the "elimination of a great amount of bookkeeping" that the "fair and simple" formula promised to entail. Over the long run the new approach shifted the nature of such fees from that of a royalty to one justified by the general cost of centralized services that AT&T provided its licensees.

AT&T's effort to reduce costs and limit its growing appetite for capital had also affected the company's relations with non-competing independents. After consultations with the Engineers Department had affirmed the technical feasibility of offering certain nonaffiliated telephone companies connections into Bell's network, the Fish administration had liberalized the conditions under which such sublicensing arrangements were made.[40] AT&T management was beginning to realize that many independents were in a better position to serve what had long been viewed as the marginal, rural market. The comparatively higher costs associated with maintaining the technical and operational standards of Bell's integrated network had squeezed profits in this segment of the marketplace. In contrast, the independents were frequently able to profitably furnish a purely local service at a price lower than that charged by the Bell telephone interests. The new arrangement, which offered noncompeting independents connections into Bell's network, enabled AT&T to indirectly establish a presence in these markets with little or no investment. What is more important, under this policy Fish had taken the first tentative steps toward focusing his company's resources and attention on the more profitable areas of its business, leaving to his potential competitors markets deemed to be less desirable from an earnings standpoint. As yet, however, the new policy on interconnection was applied on a conservative and disjointed basis. Little or no effort had been made to parlay the revised arrangements into a broad, enduring alliance with the independent movement.

In the areas of financial planning, relations with the independents, and organizational development, Fish had thus presided over a number of important innovations. As the business entered the fifth year of his presidency, its administration was more centralized and its operations were more technically integrated than had been the case just a few years before. Indeed, the company and its affiliates were now taking on the characteristics of a system, albeit one still somewhat loosely organized.

Under Fish, however, the business had not yet developed a clear sense of mission. Perhaps the most obvious manifestations of this were the absence of a consistent corporate strategy for dealing with competition; the experimental, almost ad hoc nature of the efforts to promote technical and administrative integration; and the continuing inability of management to cope with the regulatory movement that was gathering momentum throughout this period. The Bell System was not the low-cost provider of telephone service, nor had its management succeeded in attracting the majority of new customers to its premium-priced integrated services. Competition continued to thrive and to grow at a rate faster than that of the Bell interests (see app. A).

In part this failure was rooted in the unprecedented costs and technical obstacles that continued to plague the rapid integration and extension of AT&T's local and long-distance services. In part it was the result of a fundamental misunderstanding of the marketplace; many subscribers, especially residential subscribers, simply had no use for AT&T's premium long-distance offerings and therefore were inclined to purchase the less expensive local services provided by independents. In setting an agenda that began to address some of these issues and in restructuring some of the company's units into more efficient organizations, Fish had laid the groundwork for a much-needed transformation in corporate purpose and position. But he now seemed unable to push his enterprise decisively through this all-important transition. This was to be the job of his successor.

In mid-1906, quite unexpectedly, AT&T suddenly curtailed virtually all licensee capital-expenditure programs.[41] The company's prodigious efforts to confront competition head-on had severely strained its financial resources and had produced a cash-flow crisis of dangerous proportions. Although Fish assured AT&T's licensees that the emergency was only temporary, the episode underscored the fundamental inadequacies of the company's financial planning methods and indicated clearly the lack of a realistic strategy for dealing effectively with the intensifying pressures in the marketplace. The crisis undoubtedly provoked serious concern among the investment bankers represented on AT&T's board, most of whom, understandably, sought a leadership more capable of anticipating AT&T's problems. By early 1907 they had made their choice. Reaching back into the company's past for a manager of considerable experience and reputation, they appointed Theodore N. Vail as AT&T's new president and gave him the task of steering the enterprise through its troubled times.

CHAPTER 9 🐚

The Vail Years: Organizing for the Universal Network

THEODORE VAIL WAS SIXTY-ONE YEARS OLD when he became AT&T's president for the second time, in the spring of 1907. He had served on the firm's board of directors since 1902 at the behest of J. P. Morgan and his colleague George F. Baker (of First National Bank of New York), both of whom had extended their substantial holdings of AT&T stock during the years of Fish's administration.[1] Convinced that the troubled enterprise required a steady and experienced hand at the helm, the Morgan faction now charged Vail with the job of revitalizing the Bell System.

He had much to do. Strapped by burgeoning financial obligations, troubled by intense competition, haunted by a reputation as an insensitive, ruthless monopoly, and hounded by the specter of regulation and municipal ownership, the company seemed adrift, its management unable to deal effectively with the major changes taking place in the industry and American business. Over the thirteen years he spent as head of AT&T, Vail was to fashion within the Bell System—and to a considerable degree in the American public—a new consensus concerning the type of telephone system that the United States should have. He gambled that his vision of a universal, centrally managed system regulated to protect the public interest would strike a responsive chord among those frustrated with

Theodore Newton Vail, president of the AT&T Company, 1885–87 and 1907–19. Vail's second term as president of AT&T was his more important. During this period he reorganized the overall Bell System into an efficient, functionally specialized enterprise; negotiated an accommodation with the government at both the state and federal levels, which, by and large, mitigated concerns over Bell's monopoly; and strengthened the finances and the market position of his enterprise. On 15 June 1919 Vail became chairman of the board of AT&T, a post that he held until his death in April 1920.

the results of competition—that the public would choose integration over redundancy in telephone operations and regulated monopoly over a decline in the quality of service brought about by unfettered competition. Soon after taking over as head of AT&T, Vail moved to popularize his vision and implement his plans.

"One system, one management, universal service" became the credo of the Bell System under Vail. Before the public and in annual reports, which increasingly became a forum for his ideas, Vail argued that the value of a telephone system was measured by the number of subscribers that it connected together. Cooperation and operational interdependence, not competition, constituted the centerpiece of his vision. His message was clear: "Duplication of plant [was] a waste to the investor. Duplication of charges [was] a waste to the user."[2]

Regulation played a crucial role in Vail's plans. Astute enough to realize that the kind of system that he proposed—a universal, integrated monopoly—would stand little chance of gaining public approval without some form of public control, he embraced state regulation. In doing so, he broke with his company's longstanding opposition to what its management had traditionally regarded as an unwarranted intrusion on its prerogatives. But after years of unfettered competition, during which the firm's financial strengths had been sapped and its efforts to build an integrated system had been dangerously undermined, regulation became a much-preferred alternative. Vail harbored no serious objections to state regulation, provided it was "independent, intelligent, considerate, thorough and just, recognizing, as [did] the Interstate Commerce Commission . . . that capital [was] entitled to its fair return, and good management or enterprise to its reward."[3]

Not all kinds of regulation appealed to him—only state regulation, the most conservatively inclined among the lot. Regulation itself had blossomed into a movement of sweeping proportions since Vail had last managed a telephone business. Although less than a dozen states boasted a utility commission—and these were vested with varying degrees of authority over telephone rates and operations—many more states were seriously considering establishing such governing bodies.[4] It was regulation of the municipal variety that Vail hoped to outflank by accepting the authority of state utility agencies. The proponents of municipal ownership embodied a move-ment that seemed to be of a more radical bent.[5]

Vail's success in persuading a large and, as it turned out, important segment of the public to support his ideas would represent one of the more notable achievements of his administration. Also, it almost certainly would come to be regarded as one of the most masterfully orchestrated excursions in public relations of his day. But his claims for the business, and the new relationship he sought to establish between the Bell System and its public, were more than mere rhetoric. They were backed by dramatic changes in policy and outlook, in management structure and practice.

Among the areas witnessing great changes was management's attitude toward the company's independent rivals. Under Hudson and Fish the company had met the opposition with "fighting rates," patent infringement suits, and even, according to an FCC report, "propaganda campaigns" carefully designed to discredit the financial condition and operations of competitors. Even more significant was the firm's refusal to connect with most independently owned proper-ties, a policy that denied them access to AT&T's extensive intercity network. Though modified somewhat during Fish's term to permit noncompeting independents employing Western Electric equip-ment access to the Bell System, the new policy still applied only to a small number of Bell's rivals. As of 1906, only 297,000 out of an estimated 2.16 million independent stations were connected to the Bell System.[6]

Under Vail, AT&T's policy on interconnection was further liberalized. Restrictions against employing equipment manufactured by vendors other than Western Electric were dropped; in their place AT&T established certain technical standards for interconnection. The results were dramatic. In 1907 alone the number of independent stations connected to the Bell System more than doubled; in 1908 the total of nonconnecting independent stations declined for the first time (see app. A). By 1910 there were more independent telephones connected to Bell's lines than there were remaining outside the system. Three years after taking office, Vail had reversed AT&T's competitive situation in the industry.

Prompting Vail's decision to open up Bell's network to noncompeting independents were the financial strains of building the entire system alone. The same kinds of financial concerns shaped his approach to the firm's annual capital budget planning process. Vail did not want the Bell System to experience the sort of cash-flow crisis that had forced an abrupt six-month suspension in plant construction during 1906.[7] The unanticipated nature of those events had underscored the inadequacies of the company's provisional-estimate system; this weakness in the organization was brought home again the following year when a panic on Wall Street compelled Vail to impose austerities on capital spending to preserve AT&T's cash position (see app. B).[8] A firm believer in the merits of long-term financial planning, Vail introduced a new form of fiscal control, a refined version of the provisional-estimate system, with the budget cycle of 1909.[9]

Future expenditures were divided into two distinct categories: "imperative work" and "necessary work." The first included investments that would "protect the company's earning power." Necessary work involved expenditures that were important but could be deferred in times of tight money. The latter included projects that increased revenues, furnished an immediate and "proper return on the additional investment," and strengthened the existing plant.[10] Investments unlikely to pay a fair return were eliminated from the budget. In setting strict rate-of-return benchmarks for telephone company construction budgets, Vail sought to avoid the mistakes of his predecessor, who Vail thought had wasted corporate resources in areas contributing little to the future financial welfare of the enterprise.[11]

Quietly Vail centralized his administration's authority over the business's financial affairs. With the distribution of standardized provisional-estimate forms in 1908, AT&T streamlined the process of assembling, analyzing, and approving local-company budgets.[12] Four years later, AT&T engaged in its first exercise in long-range financial planning. "Foresight is certainly one of the most important factors in the proper development of the telephone system," wrote Vice President E. S. Bloom in explaining the reasons for initiating the new estimates.

> In our development studies and fundamental plans, we look ahead twenty years and carefully work out what seems in our best judg-

ment to be the broad plan which should be followed in order to satisfactorily meet the conditions of twenty years hence as we picture those conditions to-day. In our provisional estimate, we look ahead one year and lay out a program of work. In order that our work may be carried on most efficiently, we should avoid peaks as far as possible and carry out a reasonably even program of work over a period of years, thus avoiding the expense of training temporary employees, who in addition are considerably less efficient than those who have had long training and understand our methods.[13]

Although AT&T's directors were now responsible for granting final approval for Bell Company construction programs, all the work involved in assembling data and analyzing the capital needs of the business occurred at the local level. In that sense the system still was not fully centralized. The entire process depended on efficient interdepartmental planning within the operating entities and a local management both willing and able to make decisions on the basis of the new financial credo espoused by headquarters. Not everyone that sat in operating company executive suites would find this easy to do, particularly in light of their traditional autonomy in such matters. But as local capital markets diminished in relative importance as a source for funding growth and the operating companies turned increasingly to AT&T and Wall Street for financing, such discipline became a matter of course. Morgan and his colleagues in the banking industry demanded a reasonable return on their investment. Vail was determined to hold local-company executives accountable for satisfying the bankers' expectations.

How well did Vail's plan for paying more attention to the potential return on investment in evaluating capital expenditures work? The figures are mixed and perhaps a bit deceiving. Net income per share of AT&T stock dropped almost without interruption between the years 1907 and 1914, from a high of 9.18 percent to a low of 7.53 percent.[14] The years 1915 and 1916 witnessed a precipitous rise in Bell System net income per share, before the combined impact of the war and government control caused another substantial erosion in these figures. When Vail retired in 1919, AT&T was earning an 8.09 percent return on its equity. On the whole, this was not a disappointing performance, but it was not yet equivalent to the one posted by Fish in the last year of his administration.

Nevertheless, AT&T's return was far more stable than it had been. During Fish's six years in office, the company's return as a percentage of equity had vacillated between a low of 6.71 (1904) and a high of 9.18 (1906). Under Vail the spread was much narrower, most of the time falling in the range of 7.5–8.5 percent.[15] The lower but more predictable return was almost certainly preferred by many in the investment community. Indeed, although AT&T, like other large, capital-intensive enterprises, operated under significant financial constraints during this period, Vail appeared to have no difficulty in raising the funds needed to support his company's expansion. The capital that investors put into AT&T rose from about $179.6 million in 1907 to about $442.8 million in 1920. The debt advanced at a somewhat slower rate: from about $178.5 million in 1907 to $317.5 million thirteen years later.

Vail sought a balanced menu of equity and debt financing. He was convinced that "it would be manifestly unfair to the present stockholder to reduce the dividends which they have been getting and which they have been led to believe would continue." He recognized, of course, that accomplishing this objective would become increasingly difficult during a period in which "necessary concessions to the Public, and to the employees [and] other restrictions [would] tend to reduce the margin between revenue and expenses." Vail had once maintained that without increasing common stock, through debt financing he could provide the funds necessary for expansion "at a minimum rate—say, not exceeding four and five percent," and thus maintain "for a considerable period" the margins necessary to pay shareholders their anticipated dividends.[16] This expectation regarding the advantages of "leveraging" appeared to be behind his new financial strategies.

The initiatives that Vail undertook in the regulatory and financial realm during the first years of his administration set a course that would be long adhered to by succeeding generations of Bell System leaders. This was also true of the changes that he launched in the area of corporate organization.

During the last months of the Fish presidency, a special committee drawn largely from AT&T's board had met "to consider the organization of the Company and its relation to the associated companies."[17] Comprising such prominent directors and financial magnates as Senator Crane; George Baker, of the First National Bank; T. Jefferson Coolidge, of Old Colony Trust; John Waterbury, a Wall Street investor closely associated with the Morgan interests; Frederick P. Fish; and Vail himself, the committee studied in detail the bylaws and structures of other giant corporations, including U.S. Steel and several of the larger railroads.[18] These deliberations were followed by a decision to replace the company's traditional territorial management structure with a functional approach to the administration of its complex operations.

After assuming the presidency of AT&T in mid-1907, Vail moved quickly to implement the recommendations of the committee in those entities where AT&T's ownership and influence were greatest. As was so often the case, the job of organizational renovation began with Bell's long lines operation. At the beginning of 1907, before reform commenced, the intercity organization (Long Lines) had been divided into ten departments: the small headquarters staff consisted of accountants; treasurers; and attorneys; and seven operating units, including the Construction Department, the Canvassing Department, the specialized Railway Department, the Contract and Records Department, the Engineers Department, and the Maintenance and Operating Department.[19] The Construction Department and the Maintenance and Operating Department (M&O) were by far the largest Long Lines units in terms of payroll and budgets, boasting more than eight hundred employees each. M&O's operations were subdivided into six regional divisions.

These organizational arrangements had been shaped, in large measure, by the manner in which Bell's long-distance system had grown. A comparatively small Traffic Bureau (numbering only 19 employees in 1905) had handled interexchange calling forecasts; and AT&T's Construction Department, which employed a roving gang of 760 workers, had been deployed wherever Bell's long-distance subsidiary was extending or enhancing its network. Each M&O division supported those activities related to its territory in the Long Lines network, and within each division, maintenance, operations, and commercial tasks were administered on a consolidated basis at the district level. The local district, which constituted a discrete territorial unit usually no larger than an exchange, was managed by an official responsible for all nonconstruction work. This arrangement had encouraged local managers to become familiar with their unique market for telephone services and to devise specific solutions for the particular commercial and technical problems that they encountered in the areas that they served.

Vail and his colleagues, however, now considered this sort of territorial configuration an obstacle to the firm's progress in building a nationwide, interconnected system. The benefits of territorial unity no longer outweighed the advantage of functional specialization. In later explaining their decision to abandon the firm's traditional structure, Bell officials underscored their conviction that few district managers displayed either the "mental faculty or necessary training and experience to be proficient in all three branches of operation [plant, traffic, and commercial]."[20] Specialization along functional lines not only would increase the level of technical expertise at the local level but would drive management toward finding standard, system-oriented (rather than geographically focused) solutions to their problems.

Long Lines was reorganized along functional lines in the spring of 1908, following the separation of the diverse activities once combined under M&O into three, discrete units: the Plant, Traffic, and Commercial departments. A superintendent was placed in charge of each department. All three functional units were subdivided into five regional divisions, with headquarters in New York, Philadelphia, Chicago, Kansas City, and Atlanta; and these divisions were subdivided again into smaller, district-level units. Plant Department district managers, who reported up the line to division managers, handled such activities as facility and real estate maintenance, central office equipment design, and construction planning.[21] District-level managers in the Traffic Department were accountable for the efficient flow of telephone traffic, service inspection, operator services, and the establishment and maintenance of operator training programs. The Commercial Department determined toll rate schedules and devised revenue development plans, supervised Bell's advertising, maintained customer contacts, and prepared reports and recommendations dealing with the nature and extent of competition.[22]

To complement the three-column functional organization of Long Lines and to further reinforce an overall systems outlook within the business, the Vail administration realigned Western Electric and AT&T engineering activities. Not a small consideration in this reorganization was Vail's concern to achieve efficiency by eliminating redundancy in all branches of the enterprise. But the plan's primary objective was to curtail "excessive and uneconomic diversity of types of apparatus and methods" used in local and long-distance operations. "In order that this diversification may not continue," noted AT&T's chief engineer, "it will be necessary that the Licensee companies refrain from requesting the Western Electric Company to develop new types of apparatus and that they submit the supposed requirements to this Department in order that it may be determined whether the result which it is desired to obtain is a proper one and therefore should be passed on to the Western Electric Company for development."[23] Broken was the historic link between the licensees and Western Electric which in the past had frustrated Bell's programs for standardization.[24]

Standardization as a method for enhancing efficiency and cutting costs was applied elsewhere as well. After reorganization, AT&T engineers exercised complete authority over the testing and inspection standards employed by Bell's manufacturing arm.[25] Engineering focused on the establishment of systemwide standards of performance and was no longer allowed to become heavily involved in the installation and maintenance of long-distance facilities. Under the new functional arrangements, these activities were transferred to the subsidiary's new Plant Department.

More difficult to reorganize along functional lines were the activities of the associated Bell companies. Longstanding traditions of local-company autonomy represented one obstacle to such change; AT&T's lack of complete ownership control over the affiliates another. Such limitations were acknowledged when the Vail administration considered how best to implement programs of standardization and functional reorganization at the operating-company level. As Hall wrote Vail in the fall of 1909, the key to their ultimate success in this particular endeavor was a broad extension of AT&T's financial control.

> When we acquire the ownership of all the stock of any company, we are in a position for the first time, to say just how [the business]

should be handled. Up to this time, we have not been in that position with relation to any company, although we have approximated it in a few, and have now reached it in the case of the New York Telephone Company, and will soon reach it in Pennsylvania.

When our ownership is partial, and our control incomplete, we can approximate these conditions, making advances from time to time as we can, and where we show satisfactory results in the fully controlled companies, it will be possible to secure the adoption of the same methods in most of the others.[26]

Hall's point was well taken. During these years AT&T steadily increased the amount of stock it owned in its licensees. Between 1905 and 1910, AT&T increased its stake from 52 percent of outstanding telephone company stock to over 60 percent. By 1915 its equity interest in the local companies had climbed to almost 70 percent.

As the FCC has pointed out, these figures actually understate the extent of AT&T's ownership.[27] Local telephone companies were often the depositories of each other's stock, which was voted by AT&T. In 1900 the value of this pool of equity amounted to only $5.3 million, or about 3.53 percent of the total outstanding shares. By 1905, however, it had increased to $30.65 million, and by 1910 it was just over $86.5 million, or 19.75 percent of the total. While these percentages declined between 1910 and 1920, these combined holdings, both direct and indirect, gave AT&T command of 81–88 percent of licensee equity.

A third limitation on functional reorganization was that the efficiencies of specialization were primarily achieved only in Bell's larger operations. For small district operations manned by no more than a handful of employees, the straightforward adoption of the new, three-column structure would almost certainly add to the existing staff. Local managers remained skeptical concerning the thin prospects for realizing gains in economy and efficiency under these conditions. E. B. Field, president of Bell's Colorado properties, questioned the notion "that more intelligent and efficient administration could be expected from specializing work."[28] Others wanted to know how the new arrangements would affect their capacity to deal with what they thought were unique operating conditions.

Vail's concepts of order and administrative control were in many ways foreign to managers who still viewed themselves as entrepreneurs engaged in building a new enterprise. If the seminal work of Frederick W. Taylor and his disciples in scientific or systematic management enjoyed a growing appeal among AT&T executives and engineers in the East, in the hinterland their ideas played to an indifferent audience. "I, as is pretty well known," Field reminded his colleagues at AT&T,

> have always been a believer in men rather than Companies, and it is hard, therefore, for me to bring myself to ardently and consistently advocate the Three Line Organization. It seems to me that it is vital to the parent Company, as well as to every subsidiary company, to consider what kind of men we are going to educate to take the places of ourselves, who must die or drop out from natural causes. If we are bringing up a lot of people, none of whom knows the telephone business fully, it certainly looks bad to me. . . . Will it not always be true that the parent Company must vitally depend on men who are in charge locally? I would rather be building an organization that makes *man* supreme and not the Company, that is, all round intelligence, which administers the Company's affairs, and not a machine.[29]

Under previous administrations, objections such as these might have delayed or even derailed the process of reorganization, but this was no longer the case. Vail was convinced that the Bell System had to be run more efficiently if it was to survive; he was certain that functional specialization would significantly improve the management of the business. The only question that remained was the degree to which the administration of telephone operations would be centralized following functional realignment. On this issue there were two schools of thought: one pressing for a complete centralization of management, under which AT&T headquarters would assume full responsibility for the planning and performance of local operations; and another advocating a somewhat more decentralized approach.

Once again Hall seems to have made a considerable contribution to this debate. In a memo sent to Vail on the eve of local-company reorganization, he argued that proponents of a complete centralization of operations had premised their position on an assumption that he considered to be "wholly unsound," namely, "that the method of organization applicable to a small area or business [could] be expanded indefinitely" in the building of a national management structure. The functional form suffered, he said, from "very distinct limitations." If applied to the business as a whole, it would lead to the establishment of "a bureaucratic system that would be ineffective, wasteful, and in the end disasterous." Hall continued,

> In view of existing conditions, and also, in view of the fact that for legal, political and commercial reasons, we must always operate through state companies, the logical and natural way to work out the problem would be to endeavor to find an effective method of administration through subcompany organizations, controlled by a general staff.
>
> The central organization should have complete control of all policies and methods of carrying on the business, and its general staff should be made up of the best expert talent available. Through its Vice Presidents and Department heads, it should keep in close touch with every part of the field—It should be the head directing the operations of the body, and coordinating all its movements. Each state company should be in direct charge of its President, who should be held responsible for results.[30]

The operating-company presidents, Hall believed, would secure satisfactory results by adopting modern forms of organization (ensuring that their subordinates "follow the methods laid down by the general staff"), by maintaining good relations with the public, and by making sure each department carried out its functions in an efficient and economical fashion.

The division of responsibility that Hall advocated between AT&T's headquarters staff and those charged with the management of local-company operations would enable Vail and members of his administration to concentrate on the broad financial, planning, and policy issues facing the corporation. Moreover, as Hall pointed out, the kind of management structure that he proposed would furnish a suitable system of internal checks and balances, with neither the central staff nor the local-company organizations responsible for both the implementation of corporate standards and policies and the evaluation of results. It would recast the Bell System into a more effective competitor. To make this point, Hall quoted at length from an editorial appearing in the September 1909 issue of the *Century Telephone News*, an opposition journal.

> The details of [Bell's] business are so great and the territory covered so vast as to be unwiedly [*sic*] and beyond the ability of one central organization to comprehend and control—the machine of organization, no matter how ably administered, cannot intelligently pass upon its local conditions, and the Independent telephone manager, with his alertness, born of enthusiasm for his cause and the days hatred of his rival . . . will never have any difficulty in developing and controlling his territory to its utmost capacity, as against a management, such as the Bell, directed many hundreds of miles away from the source of operation.

As it turned out, the type of organization adopted was largely based on Hall's compromise. In October 1909, AT&T distributed to the licensees a twenty-one-page circular entitled *Application of Some General Principles of Organization*. The circular outlined the company's plans for a functional realignment of local operations and addressed many of the concerns raised earlier by telephone company managers. With unique local conditions in mind, Vail agreed to permit a reasonable amount of latitude in assigning certain functions. But only on a temporary basis. Ultimately, Bell officials "expected that as the organization [became] fully developed and the officers became efficient, that the arrangement of the clearly differentiated parts . . . would be almost the same in every company."[31] In the meantime, uniform specifications for local plant and traffic operations were to be established by AT&T's Engineers Department, in cooperation with local company personnel.[32]

As was the case with the territorial form of organization, the telephone company's general manager continued to exercise broad authority over the line (operating) functions of the business (see fig. 3). Staff activities, such as legal and financial affairs, were handled by the firm's president. They remained administratively separated from such operations as the construction and maintenance of telephone facilities, the installation and repair of equipment, and the leasing of instruments. The functional division of these activities took place at the level of general superintendent, one step beneath the firm's general manager, and continued down the line to the

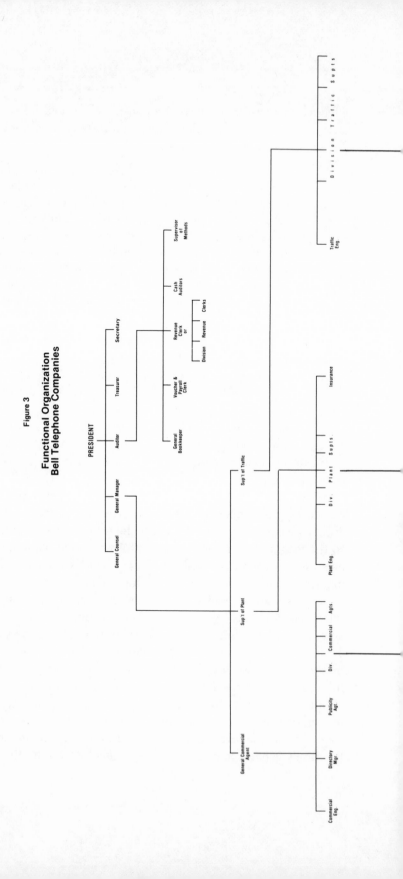

Figure 3

Functional Organization
Bell Telephone Companies

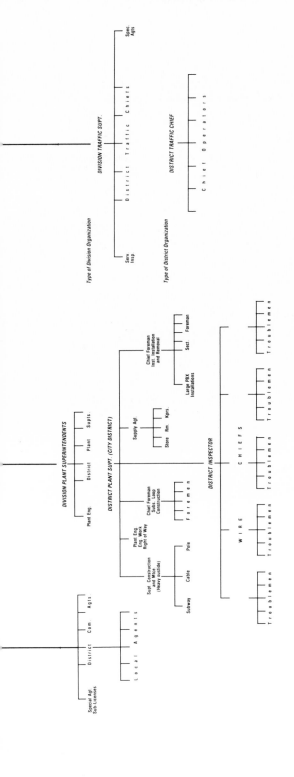

Source:
A.T.&T.Co., Historical Archives, *Application of Some General Principles of Organization.* (New York, 1909)

district level. Whereas under the territorial form of organization district managers and superintendents had been generalists, now they were specialists in either plant, traffic, or commercial work.

The distinguishing feature of the new structure was the consolidation of fundamentally related work. Under the territorial organization, subscriber canvassing had been handled by a special agent reporting directly to the general manager, while responsibility for the assignment and repair of telephones and subscriber lines (inside wiring and local circuits between the customer's residence and the central office) was in the hands of the superintendent of equipment. With the introduction of the common battery exchange system, it was no longer necessary to sustain a large staff of equipment specialists to visit subscriber premises in order to maintain telephone batteries and ensure the electrical balance of subscriber lines. In effect, such maintenance work moved into the central office, where it was now the responsibility of members of the new Plant Department. Commercial work—principally subscriber recruitment in the previous organization—expanded into the relatively new realms of advertising and relations with the independent telephone firms. The Traffic Department added new duties in the areas of traffic forecasting and engineering to its more traditional responsibilities in the area of operator training and supervision.

Changes at AT&T headquarters mirrored the functional division of responsibility established in the field, but here employees in the Plant, Traffic, and Commercial departments were not directly engaged in line operations (see fig. 4). Instead they set standards for the several Bell companies and monitored their compliance. The standards and methods that they established were enforced by Bell operating-company executives responsible for the overall earnings and operational performance of their enterprises. In each instance a degree of local discretion was retained to allow local officials the flexibility to deal with their particular market and regulatory environment. In this sense, the administration of the Bell System continued to be fairly decentralized. But even so, under the new structure, technical standardization of operations gradually spread throughout the business.

Indirectly, functional realignment allowed AT&T officials to recast operating-company boundaries along the lines of the

Figure 4

AMERICAN TELEPHONE AND TELEGRAPH COMPANY
HEADQUARTERS ORGANIZATION (FUNCTIONAL)
1912

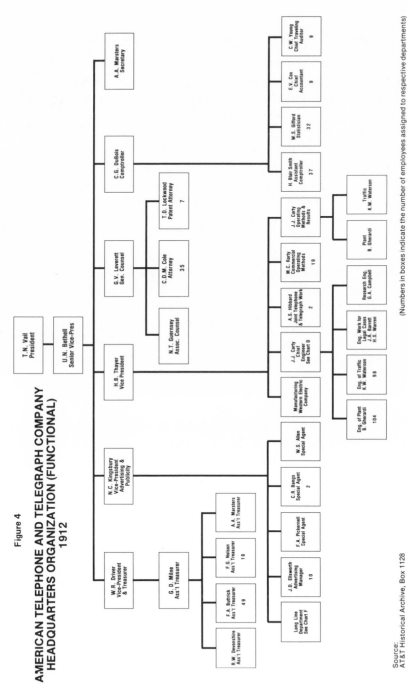

(Numbers in boxes indicate the number of employees assigned to respective departments)

Source:
AT&T Historical Archive, Box 1128

industry's emerging regulatory jurisdictions. With the new management structure driving the entire enterprise toward a national standard in technology and operations, it was no longer necessary to ensure uniformity in the administration of local service by configuring telephone company operations around what earlier officials had identified as the prevailing patterns of intercity commerce and social intercourse. The functional form put Bell officials in a position to reshape associate-company boundaries to better fit the demands of their new political environment. This was no small consideration, given that regulatory decisions were almost certain to have as much impact on earnings as competition and self-imposed cost-control measures.

The process of recasting telephone company boundaries along state lines unfolded at a gradual pace during the years of Vail's administration (1907–19) and continued well into the following decade (see maps 1 and 2). Among the first companies to undergo this kind of reorganization was the Central Union Company. Since its founding in 1883, Central Union had been managed as a multistate enterprise, with operations in Illinois, Ohio, Indiana, and a small portion of Iowa.[33] As late as 1909, plans were made to extend the scope and improve the financial viability of Central Union's operations through a consolidation with the Chicago Telephone Company and neighboring Bell companies located in the states of Michigan and Wisconsin.[34] But three years later, an abrupt change in its regulatory climate inspired a change of policy. Instead of extending its operations in new directions, the Central Union Company was eventually broken up into three smaller, statewide companies: the Indiana, Ohio, and Illinois Bell Telephone companies.

Laws that affected Central Union's rights to eminent domain and discriminated against it as a "foreign" corporation (i.e., a company incorporated in a state other than the one in which it operated and was regulated) represented one major consideration behind AT&T's decision to abandon its longstanding plans for regional consolidation. Recent legislation in Illinois, which threatened to burden the entire corporation, not just operations in that state, with a large and continuing capital-tax liability, became another.[35] But ultimately it was the new regulators who intensified pressures for a disaggregation of Central Union operations along

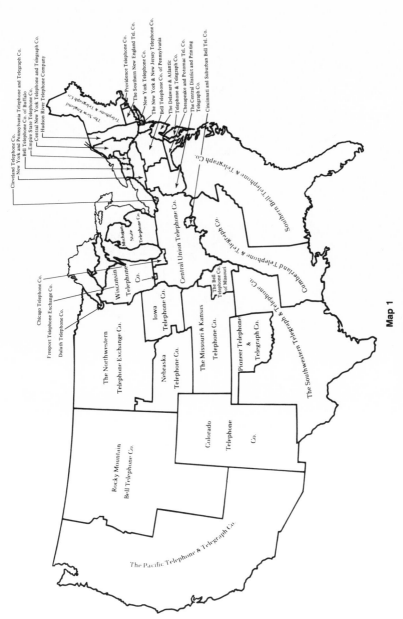

Map 1

The Bell Companies, 1893

Cleveland Telephone Co.
New York and Pennsylvania Telephone and Telegraph Co.
Bell Telephone Co. of Buffalo
Empire State Telephone Co.
Central New York Telephone and Telegraph Co.
Hudson River Telephone Company

Providence Telephone Co.
The Southern New England Tel. Co.
New York Telephone Co.
The New York & New Jersey Telephone Co.
Bell Telephone Co. of Pennsylvania
The Delaware & Atlantic Telephone & Telegraph Co.
Chesapeake and Potomac Tel. Co.
The Central District and Printing Telegraph Co.
Cincinnati and Suburban Bell Tel. Co.

The New England Telephone & Telegraph Co.

Southern Bell Telephone & Telegraph Co.

Cumberland Telephone & Telegraph Co.

Michigan State Telephone Co.

Chicago Telephone Co.
Freeport Telephone Exchange Co.
Duluth Telephone Co.

Wisconsin Telephone Co.

Central Union Telephone Co.

The Bell Telephone Co. of Missouri

The Northwestern Telephone Exchange Co.

Iowa Telephone Co.

Nebraska Telephone Co.

The Missouri & Kansas Telephone Co.

Pioneer Telephone & Telegraph Co.

The Southwestern Telegraph & Telephone Co.

Rocky Mountain Bell Telephone Co.

Colorado Telephone Co.

The Pacific Telephone & Telegraph Co.

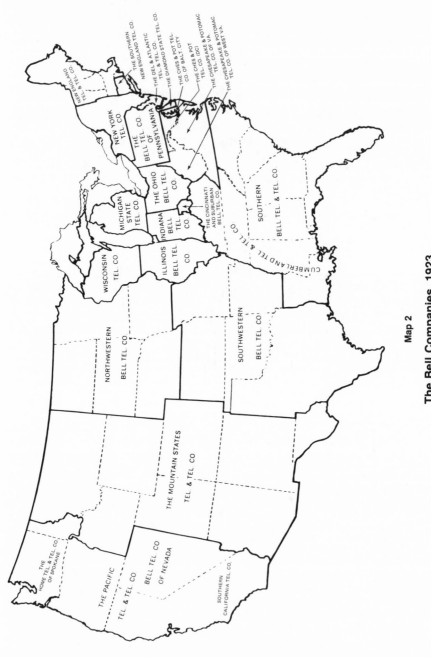

Map 2

The Bell Companies, 1923

state lines. As the president of Central Union explained to AT&T officials on two occasions in 1919: "The Indiana Public Service Commission has suggested the advisability of forming an Indiana Corporation in order that they may better supervise and regulate our operation, and the Commission is insisting that the books of the Indiana property be kept in the State of Indiana as required by law. . . . In any event, it will be necessary to form an Indiana Corporation if we are to continue to operate properties where franchises will expire in the near future."[36]

Similarly, in Ohio, public-utility officials were said to have "complained bitterly" about the fragmented, convoluted system of accounts used by the multistate Central Union enterprise and the New York–based Chesapeake and Potomac Company, which also operated in a small portion of the state.[37] Regulators sought information concerning the quality and scope of telephone operations, investments, and costs that conformed to their specific statewide jurisdictions. They employed a variety of means to encourage the Bell System to restructure its business to satisfy these demands. In the Midwest at least, Bell managers had concluded by 1920 that "under control of the State Utilities Commissions home companies seem to be in a more favorable position than Companies managed from other states."[38]

Regulatory pressures for a state-centered reconfiguration of telephone operations were also intense within the Middle Atlantic States. There politics, public agencies, and the courts had once combined to deny rate increases for the Chesapeake and Potomac multistate enterprise following its failure to segregate convincingly fixed investment and construction expenditures along state lines.[39] Plans for the consolidation of Chesapeake and Potomac, Bell Company of Pennsylvania, and New York State Telephone properties developed during the Fish administration were abandoned as it became evident that the regulators preferred to deal with statewide companies. As the chief auditor of New York Telephone pointed out in 1921, state incorporation of Bell operations "harmonize[d]" them with state regulation and "consequently tend[ed] to a simplification in our relations with public authorities."[40] As a result, these properties were gradually reorganized along state lines beginning in 1909.[41]

Not all of AT&T's multistate telephone businesses were subject to intense regulatory pressures for statewide reorganization. In the territory of what was to become the large Southwestern Telephone and Telegraph Company (Texas, Arkansas, Oklahoma, Missouri, and Kansas), it was only necessary to establish dummy state corporations to achieve equal treatment on such crucial matters as rights to eminent domain.[42] There, as was generally true throughout the South, regulation remained largely a passive institution.

On the West Coast, uncertainties surrounding the regulatory situation merited a completely different response. There an active, powerful, and frequently radical movement for "home rule" and municipal ownership vied with the emerging and more conservative state regulatory institutions for authority over the telephone industry. As early as 1900, AT&T counsel had written Pacific Telephone officials that "the liability of regulation of states through legislative action is . . . much lessened by extending the area of operation in a single Company over several states and the increased interstate character of the business. It is much more difficult to regulate a portion than the whole. The jurisdiction of the Federal Courts is also secured to a much larger extent, which I consider a great advantage, particularly in the event of being obliged to contest excessive taxation or rates fixed unreasonably low by some regulatory body."[43] For this and other reasons having to do with economy of operation and the extensive, integrated nature of Pacific Telephone's long-distance network, management decided to keep the company organized as a multistate venture.

As it turned out, Pillsbury was correct. The interstate character of AT&T's Pacific properties sheltered the enterprise from some damaging political trends through the first decade of the new century. Rulings of the Los Angeles City Council regarding Pacific Telephone earnings and rates were uniformly dismissed in state and federal judicial proceedings.[44] The movement in favor of home rule, especially strong in Los Angeles but also a worrisome trend in Seattle, Spokane, and San Francisco, gradually lost momentum as state agencies asserted their jurisdiction over telephone rates and operations.[45]

Overall, regulation in its many forms appears to have been a force making for decentralization. In several notable instances its appearance compelled Bell officials to disaggregate regional operations that had once operated as consolidated businesses. That this realignment of properties and administration created problems for Vail's universal system, at least from an engineering standpoint, was frequently evidenced in the strong opposition that such changes evoked.[46] Ironically, however, there were two sides to this phenomenon. Regulation also encouraged the consolidation of some properties operating within the same state and thus often helped Vail and his colleagues build telephone businesses that represented a more viable and more balanced configuration of urban and rural services than had their predecessors.

The best example of this was in the state of New York. From the early years of telephony, the market for telephone services in the state had been divided up among several Bell franchises. The New York City metropolitan area, including several northern New Jersey counties, fell within the domain of the Metropolitan Telephone and Telegraph Company, an enterprise jointly owned and managed by the American Bell and Western Union companies after their patent settlement in late 1879.[47] The rest of the state was licensed to companies of primarily a single-city or rural-service character, among them the Empire State Telephone and Telegraph Company, the Central New York Telephone and Telegraph Company, the Hudson River Telephone Company, the Northern New York Telephone Company, and the Troy Telephone and Telegraph Company.

Long before regulation became a factor of importance in calculations over how best to organize these businesses, Bell officials had considered statewide consolidation as a means for strengthening the earnings of their more rural-based New York operations.[48] As a report from the AT&T Engineers Department had said of Empire State's rural properties some years earlier, "The mortgage [was] too big for the farm."[49] This was undoubtedly true of Bell's other rural New York operations as well. Minority interests represented on the boards of the financially stronger companies, however, successfully resisted changes, fearing a dilution in the profit potential of their own businesses.

The appearance of municipal and state regulation forced the opposition to take a second, more favorable look at Bell's statewide revenue-balancing strategies. Vail, writing George Gould, explained that "every one of our systems that, like New York [City,] stands alone and uncombined from the revenue standpoint with the surrounding territory with which it is from a telephone business standpoint in the very closest and virtually dependent relation, has suffered in every contest with public committees, commissions, or legislative bodies, and New York will sooner or later have to meet the same question."[50]

Statewide consolidation in the long run promised to strengthen Bell's overall position in New York markets by shielding its more profitable operations from the careful and sometimes unpredictable scrutiny of regulators while permitting officials to shift financial resources around in order to best confront the challenges of competition. In the end, it eliminated the fragmented character of Bell's operations within the state and probably reinforced the trend toward statewide rate averaging. Statewide rate averaging would eventually become a distinguishing feature of Bell System subscriber charges and would be embraced by regulators as a strategy for promoting the extension of telephone service to areas of marginal earnings potential.

Unlike most of his contemporaries, Vail embraced regulation as the price of AT&T's powerful, near-monopoly position in a vital communications industry. But his success in organizing the business to deal with both the technical challenge of building an integrated, nationwide system and the emerging demands of AT&T's regulatory environment did not come without its own particular brand of judicial controversy. In 1908 AT&T acquired the Western Union Company, a move in part designed to enhance Bell's long-distance operations with the only coast-to-coast communications system then in existence. The elimination of what many regarded as the only viable alternative to Bell's long-distance service, however, was viewed in some quarters as grounds for antitrust proceedings against AT&T. Taken together with the sharp decline in competition that followed after 1911 (a decline largely fostered by Vail's aggressive policy of acquiring independent companies), this combination changed the mood of the nation's trust busters from one of passive observation to one of more determined opposition.[51]

In 1912, the U.S. Department of Justice began actively monitoring AT&T's behavior after competing independents lodged numerous protests regarding Vail's policies of acquisition and sublicensing.[52] Yet the Department of Justice seemed to be of two minds with respect to Vail's policies. Attorney General Wickersham, writing Charles A. Prouty, the chairman of the Interstate Commerce Commission, urged the latter to open a general investigation of the telephone industry in early 1913. In his letter, he argued that "the value of a telephone service depends largely upon the facility of connecting every individual telephone user with any point upon any telephone line in the United States; but this should be attained under conditions which secure to the public the most reasonable terms consistent with a fair return upon the investment and under suitable supervision and control by your honorable body."[53] Nonetheless, in July 1913, bowing to the concerns of the independents, the Department of Justice filed an antitrust suit against AT&T in the Federal District Court of Oregon. The complaint focused on AT&T's interconnection and acquisition policies and sought as redress the divestiture of the properties that Bell had recently acquired in that state.

AT&T management was under considerable pressure to negotiate a settlement. A report issued by the postmaster general on 25 November 1913 advocated government ownership of the national telephone system. From the independents came policy suggestions that would have enjoined AT&T from monopolizing the interstate telephone business, opened Bell toll lines to all companies desiring such connections, and forced the divestiture of all recently acquired independent properties that remained physically separate from Bell's network.[54] Moreover, the courts seemed inclined of late to heed the Department of Justice's interpretations of antitrust policy. Only two years before, the government had won in court victories that resulted in the divestiture of two of America's leading monopolies: Rockefeller's gigantic Standard Oil Company of New Jersey; and the American Tobacco Trust.

AT&T heeded these portents. On 19 December 1913, Nathan Kingsbury, first vice president of AT&T, signed an agreement that obligated his company to dispose of its holdings in Western Union, refrain from acquiring competing telephone companies without the Department of Justice's explicit approval, and furnish "qualified" toll connections to independent companies not yet connected to the Bell network.[55] Instead of battling the government in court, Vail chose to compromise. In effect, the company had traded away future acquisitions for government approval of Bell's existing and powerful position in the market. Interconnection sealed the bargain. Vail would not be able to provide universal service over a system wholly owned by AT&T, but his enterprise would be able to build a truly national network. The agreement, later dubbed the Kingsbury Commitment, ended telephony's second era of competition and ushered in a period of stable relations with the federal government that would last for the next twenty years.

Vail's revitalized Bell System was now characterized by uniformity of practice and policy, standardization of operations, and cooperation in planning and coordination in the commercial development of the business. The system's centrally directed, three-column functional structure would provide the firm with an organization that could achieve a high degree of technical integration without entirely sacrificing local initiative in the operating companies. As long-distance transmission became increasingly effective in the years following the perfection of the loading coil (1906), the system that emerged became the truly national network that Alexander Graham Bell had once envisioned and that Theodore N. Vail had sought to create from his first day as president of the American Telephone and Telegraph Company.

CHAPTER 10

Reflections ✍

THIS STUDY OF THE EMERGING horizontal structure of the Bell System suggests that historians of the firm will be disappointed if they expect to find uniform patterns of development or simple causal relationships in business history. Instead, they can anticipate finding evidence of complex and shifting patterns of behavior shaped by a wide variety of factors both internal and external to the firm. Frequently, individual leaders and even those beneath the top level of the firm will put their imprint on its history.

In the case of the Bell System, at least three individuals had a major impact on the company's horizontal structure in the years covered by this study. Gardiner Hubbard was a dominant force during Bell's early years, and Theodore N. Vail did more than any other single person to shape the structure and strategy of the modern telephone system. He also guided it into a new relationship with public authority at the state and federal levels. The third individual whom I would single out is E. J. Hall. Hall never headed the Bell enterprise, but from the 1880s on, Hall relentlessly promoted the types of organizational change that would provide the foundation for Vail's universal network. Hall built up within the firm the sort of organization and technological momentum that would keep the Bell System evolving toward a consolidated structure even when the company's presidents were not particularly vigorous leaders.[1]

While Hall, Vail, Hubbard, and others made decisions that had a decisive impact on the Bell System, their actions were constrained by technological, economic, and political forces beyond their control. The capital requirements and technology of the exchange business, for instance, compelled the Bell Company to alter decisively its relations with its licensees. Similarly, the development of toll services led to the administrative consolidation of contiguous exchange agencies, a process that began under the direction of parent company officials during the early 1880s. Finally, the formation of AT&T and the extension of long-distance service pressed the enterprise to coordinate and standardize the system's technology in a way that permitted the business to operate as a single, unified operation. But the responses to technological imperatives of this sort were always shaped by other considerations, such as the capital resources of the parent company; the competitive situation; and the development of a vigorous regulatory movement.

Competition in the industry was a decisive factor when Western Union was in the field and after Bell's original patents expired. The drive to protect Bell's market position influenced every choice that management made during these years. The Vail consolidation cannot be understood without considering the competition that AT&T faced and its efforts to achieve a monopoly or near-monopoly position in telephony. And the drive against competition and for centralized control cannot be understood without considering the economic characteristics of telephony and the kind of universal system that Bell officials had sought to build within the industry since the late 1880s. Vail succeeded in this regard, but he did so only by making substantial concessions to public authority.

The history of AT&T and its associated telephone enterprises suggests that those studying the evolution and behavior of modern business institutions would do well to look beyond the prevailing technological or economic imperatives of corporate development in order to understand the sometimes subtle political factors affecting company growth. Unlike the organizational development of many of the large industrial enterprises examined by Alfred D. Chandler, Jr., in his pioneering studies in the field of business history, AT&T's was dramatically influenced by the political environment.[2]

In Massachusetts, an adverse political climate limited the firm's ability to raise much-needed capital, blocking what might otherwise have been a more determined effort by Bell management to consolidate Bell's control over its licensees during the early years of exchange and toll-service expansion. Later, AT&T's drive to centralize control over the telephone system was decisively tempered by the emerging demands for state and local regulation. Eventually, during the early years of Vail's presidency, a federated structure of planning and administration, a structure particularly well suited to the jurisdictional requirements of the company's regulatory environment, was adopted. For this reason, and others, the story of the Bell System is as much a study in political economy as it is a history of how managers and engineers teamed up to build what became the world's most complex integrated communications system.

The political accommodation reached with the federal government in 1913 also had a major effect on the company's mission and on its strategic position within its industry. The Kingsbury Commitment marked the conclusion to the second competitive era and ended Vail's vigorous campaign to buy out the independents and incorporate them into the Bell System's universal network. But what is perhaps just as important, it represented a political solution to the enduring problem of monopoly—to the vexing contradictions dividing the engineer's vision of an integrated, cooperative system and the competitive notions espoused by the nation's free-market enthusiasts. Following the Kingsbury Commitment, AT&T and its affiliates built an enduring partnership with their erstwhile opposition and, with the help of state and federal agencies charged with regulating telephone service, restructured the industry into a constellation of cooperating businesses. AT&T became the nation's network manager. But it managed a network that it did not completely own. Thereafter, its attention was turned inward, and its resources were concentrated on extending the scope and quality of its own system.

What emerged from the complex evolutionary process described in the previous chapters was an unusual type of business organization. As energized by men like Hall, the Bell System was a business with a strong set of values, a corporate culture, stressing efficiency, orderly procedures, consensus, and technical expertise. As restructured under Vail's leadership, the Bell System was more centralized and technically integrated than it had been to that time. Vail created the universal network that he had promised to develop. But unlike most large firms of that time, the Bell System still left a substantial amount of authority and responsibility in the hands of the managers of the operating companies. Bell was thus an organizational hybrid and a political maverick. It was as well soon to become the largest private enterprise in the United States. By that measure, and several others, the builders of the Bell System had done their job very well.

Appendixes

APPENDIX A ❧

TOTAL TELEPHONES IN THE UNITED STATES ON 31 DECEMBER, 1876–1920

31 December	Company[a]	Private-line	Bell-owned	Service
1876	2,593	—	2,593	—
1877	9,283	—	9,283	—
1878	26,265	—	26,265	—
1879	30,872	—	30,872	—
1880[b]	47,880	—	47,880	—
1881	71,387	—	71,387	—
1882	97,728	—	97,728	—
1883	123,625	—	123,625	—
1884	134,847	12,868	147,715	—
1885	141,757	13,994	155,751	—
1886	151,012	16,121	167,133	—
1887	163,086	17,594	180,680	—
1888	175,968	18,998	194,966	—
1889	190,107	21,396	211,503	—
1890	204,597	23,260	227,857	—
1891	216,543	22,793	239,336	—
1892	232,780	28,015	260,795	—
1893	237,987	28,444	266,431	—
1894	244,162	26,219	270,381	—
1895	282,446	27,056	309,502	—
1896	325,998	28,303	354,301	—

Connecting	Subscribing	Bell Connecting	Grand Total	Nonconnecting (estimated)
—	—	—	2,593	—
—	—	—	9,283	—
—	—	—	26,265	—
—	—	—	30,872	—
—	—	—	47,880	—
—	—	—	71,387	—
—	—	—	97,728	—
—	—	—	123,625	—
—	—	—	147,715	—
—	—	—	155,751	—
—	—	—	167,133	—
—	—	—	180,680	—
—	—	—	194,966	—
—	—	—	211,503	—
—	—	—	227,857	—
—	—	—	239,336	—
—	—	—	260,795	—
—	—	—	266,431	—
—	—	—	270,381	15,000
—	—	—	309,502	30,000
—	—	—	354,301	50,000

APPENDIX A—CONTINUED

31 December	Company[a]	Private-line	Bell-owned	Service
1897	385,058	30,155	415,213	—
1898	466,077	29,721	495,798	—
1899	633,918	32,815	666,733	—
1900	801,947	33,964	835,911	—
1901	1,021,820	39,292	1,061,112	—
1902	1,279,241	37,937	1,317,178	—
1903	1,526,957	36,984	1,563,941	—
1904	1,801,667	36,367	1,838,034	—
1905	2,243,576	41,011	2,284,587	—
1906	2,729,116	44,431	2,773,547	—
1907	2,965,546	46,965	3,012,511	71,173
1908	3,130,154	46,240	3,176,394	85,091
1909	3,476,424	45,655	3,522,079	111,823
1910	3,887,690	45,366	3,933,056	142,978
1911	4,306,320	45,517	4,351,837	167,851
1912	4,757,071	46,732	4,803,803	196,376
1913	5,206,791	48,017	5,254,808	208,418
1914	5,535,857	48,996	5,584,853	227,151
1915	5,919,462	48,648	5,968,110	236,438
1916	6,494,591	50,899	6,545,490	250,436
1917	6,978,396	53,134	7,031,530	270,311
1918	7,148,722	53,035	7,201,757	291,906
1919	7,687,219	51,940	7,739,159	296,289
1920	8,277,600	56,379	8,333,979	299,714

Source: AT&T comptroller's annual report for 1934, pt. 1, statement no. 45.

[a]Prior to 1907, company stations included "service stations."

[b]First official report.

Connecting	Subscribing	Bell Connecting	Grand Total	Nonconnecting (estimated)
—	—	—	415,213	100,000
—	—	—	495,798	185,000
10,000	—	10,000	676,733	328,000
20,000	—	20,000	855,911	500,000
47,961	—	47,961	1,109,073	692,000
84,021	—	84,021	1,401,199	969,845
120,936	—	120,936	1,684,877	1,124,000
167,213	—	167,213	2,005,247	1,348,000
246,337	—	246,337	2,530,924	1,596,000
297,218	—	297,218	3,070,765	1,862,000
755,316	—	826,489	3,839,000	2,279,578
1,103,144	—	1,188,235	4,364,629	2,119,000
1,508,790	—	1,620,613	5,142,692	1,853,000
1,806,685	—	1,949,663	5,882,719	1,752,648
2,112,937	—	2,280,788	6,632,625	1,716,093
2,455,895	—	2,652,271	7,436,074	1,518,862
2,669,791	—	2,878,209	8,133,017	1,409,492
2,836,989	9,619	3,073,759	8,658,612	1,387,806
2,946,673	21,274	3,204,385	9,172,495	1,351,002
3,051,266	46,418	3,348,120	9,893,610	1,347,822
3,173,837	63,898	3,508,046	10,539,576	1,244,066
3,498,662	73,674	3,864,242	11,065,999	1,011,638
3,685,372	74,927	4,056,588	11,795,747	872,727
3,885,658	82,584	4,267,956	12,601,935	809,444

APPENDIX B 🖊

EXPANSION IN TELEPHONE PLANT WITHIN THE BELL SYSTEM, 1885–1912

Year	Total Telephone Plant	Increase	Percent Increase
1885	$ 38,618,600	—	—
1886	38,325,431	$ 293,169	0.759%
1887	40,799,143	2,473,712	6.454
1888	44,436,342	3,637,199	8.915
1889	51,572,129	7,135,787	16.058
1890	58,512,400	6,940,271	13.457
1891	62,190,195	3,677,795	6.285
1892	67,635,701	5,445,506	8.756
1893	73,136,242	5,500,541	8.133
1894	77,731,161	4,594,919	6.283
1895	87,858,500	10,127,339	13.029
1896	95,241,646	7,383,146	8.403
1897	104,487,524	9,245,878	9.708
1898	118,123,841	13,636,317	13.051
1899	145,511,290	27,387,449	23.185
1900	180,699,800	35,188,510	14.183
1901	211,780,200	31,080,400	17.200
1902	250,013,200	38,233,000	18.053
1903	284,567,800	34,554,600	13.815
1904	316,520,600	31,952,800	11.229

Year	Total Telephone Plant	Increase	Percent Increase
1905	$368,065,300	$51,544,700	16.285%
1906	450,061,400	81,996,100	22.278
1907	502,987,900	52,926,500	10.522
1908	528,717,000	25,729,100	5.115
1909	557,417,146	28,700,146	5.428
1910	610,999,964	53,582,818	9.613
1911	666,660,702	55,660,738	9.110
1912	742,287,631	75,626,929	11.344

Source: FCC, Telephone Investigation, 138.

APPENDIX C

NET INCOME PER SHARE OF CAPITAL STOCK

AMERICAN TELEPHONE AND TELEGRAPH COMPANY

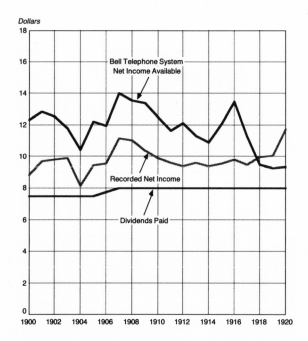

Source: FCC, Telephone Investigation, p.507

APPENDIX D 🙐

BELL SYSTEM
Principal Balance Sheet Data and Relative Size of Such Data
at Five-Year Intervals

As of December 31, 1885 to 1920 , Inclusive

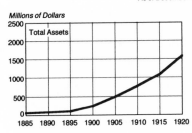

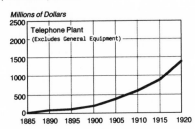

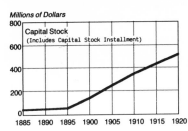

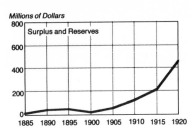

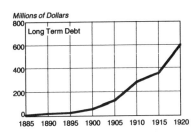

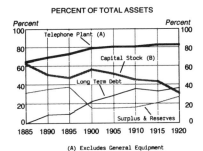

(A) Excludes General Equipment
(B) Includes Capital Stock Installments

Source: FCC, Telephone Investigation, p.40

Appendix E ❧

Growth of Bell System as Reflected by Plant and Operating Statistics at Five-Year Intervals

As of December 31, 1885 to 1920 , Inclusive

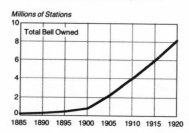

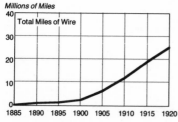

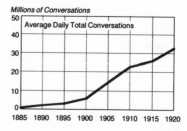

(A) Does Not Include the Employees of Western Electric Co., Inc.

(B) Does Not Include Long Lines Employees

Source: FCC, Telephone Investigation, p.42

(C) Published Figure, 14,517

Notes ❦

Chapter 1   The Great Yankee Invention

The title of this chapter is taken from remarks recorded by visitors to the Bell Telephone Company's exhibition at the American Institute Fair, convened in New York City in October 1877 (Telephone Company of New York, *Visitors Book*, 4 October 1877, American Telephone and Telegraph Company Archives, 195 Broadway, New York, N.Y. [hereafter cited as AT&T Archives]).

1. Two books that place this scholarship in a broad intellectual and political context are John Higham et al., *History: Professional Scholarship in America* (Englewood Cliffs, N.J.: Prentice Hall, 1965); and Richard Hofstadter, *The Progressive Historians: Turner, Beard, Parrington* (New York: Alfred A. Knopf, 1968). For examples of this trend in historical scholarship see idem, *The Age of Reform: From Bryan to F.D.R.* (New York: Alfred A. Knopf, 1955); Gabriel Kolko, *The Triumph of Conservatism: A Reinterpretation of American History, 1900–1916* (1963; reprint ed., Chicago: Quadrangle Paperback, 1967); and Robert H. Wiebe, *The Search for Order: 1877–1920*, The Making of America, ed. David Donald (New York: Hill & Wang, 1967).

2. John R. Pierce, "The Telephone and Society in the Past 100 Years," in *The Social Impact of the Telephone*, ed. Ithiel de Sola Pool (Cambridge, Mass.: MIT Press, 1977), 159–95; Jean Gottman, "Megalopolis and Antipolis: The Telephone and the Structure of the City," ibid., 303–17; Ronald Abler, "The Telephone and the Evolution of the American Metropolitan System," ibid., 318–41; J. Alan Moyer, "Urban Growth and the Development of the Telephone: Some Relationships at the Turn of the Century," ibid., 342–69. See also John Brooks, *Telephone: The First Hundred Years* (New York: Harper & Row, 1975), 7–10.

3. Catherine MacKenzie, *Alexander Graham Bell: The Man Who Contracted Space* (New York: Grosset & Dunlap, 1928); Albert Bigelow Paine, *Theodore N. Vail: A Biography* (New York and London: Harper & Brothers, 1921).

4. Robert V. Bruce, *Bell: Alexander Graham Bell and the Conquest of Solitude* (Boston and Toronto: Little Brown & Co., 1973).

5. Arthur S. Pier, *Forbes: Telephone Pioneer* (New York: Dodd, Mead & Co., 1953).

6. J. Warren Stehman, *The Financial History of the American Telephone and Telegraph Company* (Boston and New York: Houghton Mifflin Co., 1925).

7. Ibid., 51, 292.

8. Ellis W. Hawley, *The New Deal and the Problem of Monopoly: A Study in Economic Ambivalence* (Princeton: Princeton University Press, 1966), 383–456.

9. Federal Communications Commission, *Investigation of the Telephone Industry in the United States* (Washington, D.C.: U.S. Government Printing Office, 1939); hereinafter cited as FCC, *Telephone Investigation.*

10. Horace Coon, *American Tel & Tel: The Story of a Great Monopoly* (New York and Toronto: Longmans, Green & Co., 1939; reprint ed., Freeport, N.Y.: Books for Libraries Press, 1971); N. R. Danielian, *AT&T: The Story of Industrial Conquest* (New York: Vanguard Press, 1939); Matthew Josephson, *The Robber Barons: The Great American Capitalists—1861–1901* (New York: Harcourt, Brace & World, 1934).

11. Danielian, *AT&T,* 71–74, 379.

12. Stehman, *Financial History,* 42; Danielian, *AT&T,* 58, 379.

13. Danielian, *AT&T,* 349; Coon, *Great Monopoly,* 2.

14. Danielian, *AT&T,* 380.

15. This concern surfaced in several forums during this period and was best captured in the work of Sinclair Lewis (see Sinclair Lewis, *It Can't Happen Here* [1935; reprint ed., New York: New American Library, 1963]).

16. Brooks, *Telephone.*

17. Alfred D. Chandler, Jr., *Strategy and Structure: Chapters in the History of the American Industrial Enterprise* (Cambridge, Mass.: MIT Press, 1962); Alfred D. Chandler, Jr., *The Visible Hand: The Managerial Revolution in American Business* (Cambridge, Mass.: Harvard University Press, 1977).

18. Chandler, *Visible Hand,* 89.

19. Three basic functions are involved in the provision of goods and

services: the acquisition of raw materials, the production of a good or service from such materials, and the distribution of a final product. A firm that performs two or three of these activities itself is said to be vertically integrated. The Bell System, for example, was a vertically integrated enterprise, combining within a single business the functions of purchasing raw materials, manufacturing (Western Electric Company), and distribution (the Bell Telephone companies and AT&T Long Lines). All of the parts of the business performing a similar function (e.g., distribution) are said to be horizontally related to each other, and that relationship, marked by lines of corporate oversight and authority, defines the horizontal structure of the business. Thus, the corporate ties encompassing the management of the twenty-three local Bell Telephone companies that together delivered local exchange services across the country and AT&T represent the horizontal structure of the business.

20. Such political factors included state laws and the state legislative processes affecting the incorporation, capitalization, and structure of the Bell enterprise between 1880 and 1899. After 1899, Bell's emerging regulatory environment exercised a considerable influence on the shape of the firm.

Chapter 2   Humble Origins

1. In his initial efforts to design a multiple telegraph, Bell regulated the electrical resistance between two wires by dipping their respective ends into a vessel of water, controlling the resistance by adjusting the depth to which the wires were submerged. He later used the same principle in improving the method for producing a variable resistance in his telephones. His early telephones employing this concept were therefore designated "liquid" transmitters (Frederick Leland Rhodes, *Beginnings of Telephony* [New York: Harper & Brothers, 1929], 27–28).

2. Thomas A. Watson, *The Birth and Babyhood of the Telephone* (New York: AT&T, 1913), 24.

3. U.S. Congress, Senate, Committee on Education and Labor, *Report upon the Relations between Labor and Capital*, 48th Cong. (Washington, D.C.: U.S. Government Printing Office, 1885), 882; Thomas A. Watson, *Exploring Life* (New York: D. Appleton & Co., 1926), 107.

4. Thomas B. Doolittle, 3 August 1906, AT&T Archives, box 1057.

5. *New Haven Daily Morning Journal and Courier*, 28 April 1877.

6. *Sunday Herald,* 22 October 1876; *Boston Times,* 10 December 1876; *Lowell Citizen,* 25 April 1877.
7. Alexander Graham Bell's early association with Gardiner Hubbard and Thomas Sanders is chronicled in Bruce, *Bell,* 49–143. For a popular account of these events see Brooks, *Telephone,* 40–55.
8. The board of managers comprised Gardiner Hubbard, Bell, Sanders, Watson, and Charles Hubbard (FCC, *Telephone Investigation,* 4).
9. The shares were distributed as follows: to Gardiner G. Hubbard, 1,387; to Gertrude M. Hubbard, wife of Gardiner Hubbard, 100; to Mabel G. Bell, daughter of Gardiner Hubbard and wife of Alexander Graham Bell, 1,497; to Thomas Sanders, 1,497; to Thomas Watson, 499; to Charles E. Hubbard, 10; and to Alexander Graham Bell, 10 (FCC, *Telephone Investigation,* 4).
10. Brooks, *Telephone,* 55.
11. "The Telephone," May 1877, AT&T Archives, box 1097; Gardiner G. Hubbard to board of managers, July 1877, ibid., box 1001.
12. Rosario J. Tosiello, "The Birth and Early Years of the Bell Telephone System, 1876–1880" (Ph.D. diss., Boston University, 1971), 84–90.
13. Hubbard to Thomas Watson, 11 April 1878, AT&T Archives, box 1173. The actual revenue stream produced for the company by Hubbard's leasing and pricing strategies is difficult to estimate with any confidence. The Bell Company's first licensed manufacturer, Charles Williams, charged the firm between $2.29 and $2.86 per telephone instrument and between $8.00 and $12.00 for the call bells used in signaling. These charges did not include the cost of instrument inspection, an expense that the Bell Company assumed; nor did they cover the expense of distribution and marketing, a substantial portion of which was later shifted to the licensees. Overall, the company's prices appear to have been designed to recover the expense of instrument production and distribution quickly, perhaps even within the span of one year. But the income generated from such arrangements alone proved to be inadequate for underwriting the rapid growth in manufacturing expenses associated with expanding demand and production during this period. For a more detailed treatment of this problem see George D. Smith, *The Anatomy of a Business Strategy: Bell, Western Electric, and the Origins of the American Telephone Industry* (Baltimore: Johns Hopkins University Press, 1985), chap. 3.
14. Brooks, *Telephone,* 76–81.
15. The FCC noted that from the fall of 1877 until the original Bell patents expired in 1893, the Bell Company and its successors filed

over six hundred patent infringement suits. In most instances, the filing alone was enough to discourage competition. Only a relatively small number of suits were prosecuted, and just five reached the Supreme Court before the patents expired (FCC, *Telephone Investigation*, 125).

16. "The Telephone."
17. Hubbard to board of managers, July 1877, 1–4, AT&T Archives, box 1001.
18. Correspondence describing the negotiations leading up to a license between the Bell Company and Stearns and George is contained in a folder entitled "Stearns and George—Agents—Massachusetts and New Hampshire—1877–1880," ibid., box 1185.
19. Correspondence describing the license granted to George Coy is contained in a folder entitled "Wayward, William B.—Agents (Western) Connecticut and (Western) Massachusetts, 1877–1878," ibid., box 1190.
20. A record of the Bell Company's early licensees, the terms of their contracts, and any amendments subsequently made in the contracts is contained in American Bell Record Book, vol. 1, AT&T Archives.
21. Deposition of Watson in J. J. Storrow, comp., *Proofs by and about Alexander Graham Bell, 1875–1889*, 2 vols. (n.p., n.d.), 2:262, in ibid.
22. Hubbard to board of managers, July 1877, ibid., box 1001.
23. Tosiello, "Birth and Early Years," 146.
24. Hubbard to Thomas Sanders, 20 April 1878, AT&T Archives, box 1003.
25. Hubbard to George L. Bradley, 15 December 1877, ibid., box 1001.
26. Ibid.
27. Sanders to Hubbard, 16 December 1877; Bradley to Sanders, 18 December 1877, both ibid.
28. Hubbard to Sanders, 26 December 1877, ibid.
29. Hubbard to Bradley, 24 October 1877, General Manager Letter Books, vols. 11–177, AT&T Archives (hereafter cited as GMLB), 11:323.
30. FCC, *Telephone Investigation*, 4.
31. Hubbard to Bradley, 27 October 1877, GMLB 11:323.
32. E. T. Holmes, *A Wonderful Fifty Years* (privately published, 1917), 59.
33. "E. T. Holmes to My Father," 11 May 1877, AT&T Archives, box 1057.
34. Doolittle to Thomas D. Lockwood, 18 January 1894, ibid.
35. Statement of Doolittle, 6 June 1882, ibid.
36. J. Leigh Walsh, *Connecticut Pioneers in Telephony: The Origins and*

Growth of the Telephone Industry in Connecticut (New Haven: Telephone Pioneers of America, 1950), 44, 53–55; C. H. Gilbert, "Development in Communications," 2, AT&T Communications Records Center (Historical File), 33 Thomas Street, New York, N.Y.

37. Walsh, *Connecticut Pioneers*, 55, 90.

38. "E. T. Holmes to My Father."

39. Holmes to Hubbard, 19 July 1877, and Hubbard to Holmes, 10 August 1877, both AT&T Archives, box 1057.

40. Watson to Mr. Durant, 26 March 1878, GMLB 13:172; Watson to Mr. Gleason, 27 March 1878, ibid., 186–87. Sanders wrote a licensee, "Your conception of the District business seems to be entirely wrong. . . . It is a new business to all of us, and different systems work successfully in different localities, but your idea of a separate wire for each customer will not work" (Sanders to Mr. Heald, 26 March 1878, ibid., 174).

41. Bell Telephone Company, "Instructions to Agents No. 3," 1 February 1878, AT&T Archives, box 1001.

42. Bell Company to Charles H. Haskins, 10 August 1878, GMLB 15:51–52.

43. Watson to Hubbard, 26 March 1878, ibid., 13:167–68; Watson to Mr. Bofinger, 26 March 1878, ibid., 170–71; Watson to Bell, 30 August 1878, ibid., 15:123–216.

44. Watson to Frederick A. Gower, 29 August 1878, ibid., vol. 15.

45. FCC, *Telephone Investigation*, 123–25.

46. Notebook of Thomas Watson, 2 March 1878, AT&T Archives, box 1069; Hubbard to Sanders, 22 May 1878, ibid., box 1152.

47. Sanders to Hubbard, 5 March 1878, GMLB 13:18–20.

48. Thomas E. Cornish to Hubbard, 9 March 1878, AT&T Archives, box 1176.

49. Sanders to Hubbard, 12 April 1878, ibid., box 1003.

50. The original Bell Company had not been formally incorporated within the state of Massachusetts, an unnecessary procedure given its organization as a limited partnership. As the company grew, and in particular as the need to borrow funds from outside investors arose, Boston's financial community insisted on incorporation as the price of their financial support.

51. The June 29th Association included: Gardiner G. Hubbard, Thomas Sanders, Charles E. Hubbard, Thomas A. Watson, A. O. Morgan, Alexander Cochrane, George L. Bradley, T. B. Bailey, James Sturgis, and Joseph Goodspeed.

52. Albert Bigelow Paine, *In One Man's Life* (New York and London: Harper & Brothers, 1921), 108–9.

Chapter 3   The Uncertain Years

1. Paine, *Vail*, 2–18.
2. William D. Baldwin to Dr. Samuel White, 11 December 1879, AT&T Archives, box 1110. On the legal history of the Dowd case see Brooks, *Telephone*, 69–70.
3. Charles A. Cheever to Sanders, 4 February 1878, and Cheever to Hubbard, 25 June and 25 January 1878, all AT&T Archives, box 1024. On the original arrangements establishing Bell's New York operations see Agreement between the Telephone Company of New York and Charles A. Cheever and Milborn L. Roosevelt, 18 October 1877, ibid.
4. Hubbard to E. M. Barton, 10 June 1878, ibid., box 1156.
5. Theodore N. Vail to Bradley, 14 November 1878, GMLB 16:336–40. Although such defections were few, they were nonetheless troubling. By late 1879, Western Union had acquired complete control of the Bell Company's Bridgeport, Connecticut, and Columbus, Ohio, exchanges and held a majority interest in licensees located in St. Louis and in Providence and Pawtucket, Rhode Island. For a complete listing of Western Union's telephone company holdings see "Contract of November 10, 1878," schedule C, AT&T Archives, box 1006.
6. Tosiello, "Birth and Early Years," 133–35.
7. Sanders to Eldred, 31 August 1878, GMLB 15:128.
8. Hubbard to Messrs. Patrick and Carter, 4 October 1878, ibid., 394.
9. Western Union's growing presence in the field understandably discouraged local investors from embracing Bell's cause, particularly in Chicago and New York. Hubbard found it impossible to enlist a replacement for Anson Stager (Chicago) willing to underwrite exchange expansion on the scale required to compete with Western Union in its own backyard (Sanders to Hubbard, 24 August 1878, ibid., 97).
10. Hubbard to Vail, 24 October 1878, ibid., 417.
11. Hubbard to G. W. Balch, 4 October 1878, ibid., 399–400.
12. Vail to Oscar Madden, 26 November 1878, ibid., 16:464–66.
13. In mid-December Vail informed his assistant, general superintendent of agencies Oscar Madden, that Bradley and Sanders had agreed to the details of the new contracts (Vail to Madden, 16 December 1878, ibid., 17:202).
14. Hubbard to the stockholders of the Bell Telephone Company, 8 February 1879, AT&T Archives, box 1003, 6, 8, 1; see also Hubbard to Alexander Graham Bell, 28 February 1879, ibid., box 1003.

15. Hubbard to the stockholders of the Bell Telephone Company, 8 February 1879, ibid., 3, 11.
16. Ibid., 9.
17. Tosiello, "Birth and Early Years," 368–72.
18. Thomas Sanders, "Report to Stockholders of the New England Telephone Company," 27 January 1879, AT&T Archives, box 1003.
19. William H. Forbes to Bell, 5 March 1879, ibid.
20. Minutes of Meeting of the Directors of the Bell Telephone Company, 23 January 1879, ibid., box 1001.
21. Hubbard to Madden, 22 January 1879, ibid., box 1197.
22. Forbes to Bell, 5 March 1879, ibid., box 1003.
23. Pier, Forbes, chap. 7.
24. Forbes to Bell, 5 March 1879, AT&T Archives, box 1003.
25. Hubbard to Forbes, 26 February 1879, ibid.
26. Sanders to Forbes, 11 March 1879, ibid.
27. Alexander Graham Bell to the Directors of the National Bell Telephone Company, 11 March 1879, ibid.
28. National Bell Telephone Company, Agreement of Association, 17 February 1879, ibid., box 1008.
29. National Bell Telephone Company, Certificate of Incorporation (copy), 13 March 1878, ibid.
30. Minutes of Meeting of the Stockholders of the National Bell Telephone Company, 10 March 1879, AT&T Secretary's Office, 550 Madison Avenue, New York, N.Y.
31. Minutes of Meeting of the Directors of the National Bell Telephone Company, 11 March 1879, AT&T Secretary's Office.

Chapter 4   Accommodation with Western Union

1. James D. Reid, The Telegraph in America: Its Founders, Promoters and Noted Men (New York: John Polhemus, Publisher, 1886), 518–39, 563–66; Chandler, Visible Hand, 89.
2. Gerald W. Brock, The Telecommunications Industry: The Dynamics of Market Structure (Cambridge, Mass.: Harvard University Press, 1981), 83–87, 91.
3. Cheever to Hubbard, 11 November 1877, AT&T Archives, box 1001.
4. George Walker to Hubbard, 25 January 1878, ibid., box 1176.
5. Sanders to Hubbard, 30 January 1878, GMLB 12:285–86; Hubbard to Cheever, 4 February 1878, AT&T Archives, box 1006.

6. Walker to Hubbard, 25 January 1878, ibid., box 1176.
7. Vail to E. J. Hall, Jr., 12 December 1878, GMLB 17:160–61; Vail to Madden, 10 July 1879, ibid., 25:238–40.
8. Hubbard to the Executive Committee of the Bell Telephone Company, August 1878, AT&T Archives, box 1164; Vail to Bradley, 12 November 1878, ibid., box 1131; Hall to Vail, 22 February 1879, ibid., box 1164; Vail to Watson, 4 March 1879, ibid., box 1194; Vail to Doolittle, 23 April 1879, GMLB 21:492–93.
9. White to Forbes, 11 April 1879, "Letters about Western Union Settlement—1879" (hereafter cited as "Letters") (copy, 1924), AT&T Archives.
10. Ibid.
11. Forbes to White, 18 April 1879, ibid.
12. White to Forbes, 19, 22 April 1879; Forbes to White, 30 April 1879; and White to Forbes and Bradley, 14 May 1879, all in ibid.
13. White to Forbes, 3 May 1879, ibid. The portion of the letter quoted here is quoted verbatim in Forbes's reply requesting clarification of these terms (Forbes to White, 19 May 1879, Presidents' Letter Books, 43 vols., AT&T Archives [1879–1930], 1G:44).
14. Forbes to White, 6 June 1879; and White to Forbes, 11 June 1879, both in "Letters."
15. Novrin Green to Forbes, 2 July 1879, ibid.
16. Ormes to Forbes, 28, 29 July 1879, AT&T Archives, box 1034.
17. D. H. Lauderback to Forbes, 26 July 1879, ibid.
18. Vail to Ormes, 27 December 1878, ibid.
19. Minutes of Meeting of the Directors of the National Bell Telephone Company, June 1879, AT&T Secretary's Office.
20. White to Forbes, 21 May 1879, "Letters."
21. In a few instances, Madden was given permission to imply that the Forbes administration would not stand in the way of a timely reconciliation of local differences between Western Union and the licensees as long as the final arrangements adhered strictly to the principles embodied in the Ormes pact (Forbes to Madden, 28 August 1879, Presidents' Letter Books, 1G:91–92.
22. Vail to Forbes, 30 July 1879, and Chauncey Smith to Forbes, 5 August 1879, both AT&T Archives, box 1034; Forbes to Smith, 4 August 1879, Presidents' Letter Books, 1G:86–88.
23. The transfer of properties once owned and managed by the telegraph interests took place over the next two years and was in many cases delayed owing to the fact that Bell officials had compiled relatively little information on Western Union operations.

24. The Western Union Contract, 10 November 1879, schedule 3, app., AT&T Archives, box 1006.
25. Forbes to Smith, 4 August 1879, Presidents' Letter Books, 1G:86; Smith to Forbes, 4 August 1879, AT&T Archives, box 1006.
26. Western Union Contract, p. 1.
27. Smith to Forbes, 21 October 1879, AT&T Archives, box 1034.

Chapter 5   Building the Foundations for the Modern Bell System

1. Theodore N. Vail, "Report on Operations of the Telephone Business," 19 March 1880, 1, 4, AT&T Archives, box 1080. Ten thousand telephones remained in inventory, and a comparatively small number—about nineteen hundred—were exported to Europe. Vail indicated that his "highest estimate" of the demand for telephones originally had been one subscriber for every one hundred residents.
2. Ibid., 1–3. Instrument requirements of the new licensees, in Vail's opinion, were not reflected in the first-quarter statistics, since local agents usually placed their orders for telephones after their circuits were erected and subscribers enlisted (this in order to avoid paying a rental on idle inventory).
3. Ibid., 11.
4. The precise point at which diseconomies of scale began to take place was not recognized until 1887, although the existence of such a point was generally acknowledged to be between one hundred and one thousand subscribers as early as 1881 (*The Second Meeting of the National Telephone Exchange Association, April 5–6 1880* [New Haven: Tuttle, Morehouse, & Taylor, Printers, 1881], 137–39).
5. Vail, "Report," 4.
6. "Report of the Committee Appointed to Consider What Steps are Necessary to Prepare for the Work Now before the Company," 1879, AT&T Archives, box 1195.
7. Ibid., 1.
8. In terms of capitalization alone, the new enterprise would be launched on a much grander scale than that of its predecessor, which had been incorporated in 1879 with an $850,000 ceiling on its financial resources.
9. Stehman, *Financial History*, 19. On 11 June 1879 National Bell stock had sold for $110.50. The last five hundred shares of the company's stock sold at $600 per share in the fall of 1879 (see folder entitled "List

of Stock Prices—June 11, 1879–December 28, 1880," AT&T Archives, box 1050).

10. J. J. Storrow, "Points of Good Service," January 1880, 3–5, ibid., box 1326.

11. Ibid., 11; emphasis in the original.

12. Forbes to J. Q. A. Brackett, 2 February 1880, Presidents' Letter Books, 1G:178; Commonwealth of Massachusetts, Senate, Committee on Mercantile Affairs, *An Act to Incorporate the American Bell Telephone Company, no. 23 (22 January 1880)*, AT&T Archives, box 1007.

13. Storrow, "Points of Good Service," 4.

14. Forbes to Green, 6 February 1880, Presidents' Letter Books, 1G:185–86.

15. Commonwealth of Massachusetts, *An Act to Incorporate the American Bell Telephone Company (2 April 1880)*, AT&T Archives, box 1007.

16. Ibid.

17. American Bell Record Book, vol. 1, AT&T Archives. At the time that National Bell began issuing standard licensing agreements, its management adopted a numbering scheme that would distinguish the kind of service or business that agents were allowed to develop under contract. This tradition continued well into the 1880s under the auspices of American Bell, even though as consolidated telephone companies, the licensees came to enjoy rights to enter a broad range of exchange- and toll-related businesses. The major categories in this licensing scheme were as follows: forms 109, 109B, and 109D for exchange service; 113, 113B, 113C, and 113D for the extraterritorial business; and 116 for private-line operations. The letters *B, C,* and *D* refer to general amendments incorporated into the standard agreements.

18. Vail to Forbes, 17 March 1880, GMLB 14:407–8.

19. Exchange Contract, forms 109B and 109C, AT&T Archives, box 1008. Notable exceptions to this rule included the nine-year exchange contract that American Bell granted E. J. Hall, Jr., for Buffalo and surrounding counties (Contract no. 223, Bell Telephone Company of Buffalo, 17 January 1880), and the "permanent" contracts issued to those exchanges in which the company held an equity interest: New York City, Philadelphia, Chicago, and Boston (American Bell Record Book, vol. 1, AT&T Archives).

20. *Report on the Proceedings of the National Exchange Convention Held at Niagara Falls, New York, September 7–10, 1880* (New Haven: Tuttle, Morehouse, & Taylor, Printers, 1881) (hereafter cited as *Exchange Association Report*), 87.

21. Ibid.

22. Exchange Contract, forms 109, 109B, and 109C, AT&T Archives, box 1008.
23. Lockwood to Frederick P. Fish, 10 April 1906, ibid., box 1183. This letter to Fish includes Lockwood's early recollection of these events. Before mid-1879, toll-line operations rarely exceeded a distance of seven miles.
24. Ibid.
25. Edward F. Coburn to Bradley, 9 September 1879, ibid., box 1056; W. O. Witcome to the American Bell Telephone Company, 20 May 1880, ibid., box 1210; R. W. Devonshire to Madden, memo on the Boston to Lowell Line, ibid.
26. Charles J. Glidden to Madden, 18 May 1880, ibid., box 1210; Vail to J. N. George, 29 May 1880, GMLB 48:412. Glidden had calculated that on anticipated total first-year revenues of $2,400, connecting companies would earn an "originating commission" of about $180, an amount that would have "to pay [for] an operator and [cover] the cost of collecting bills from 125 to 200 subscribers."
27. F. B. Knight to the American Bell Telephone Company, 23 April 1882, cited in Roger B. Hill, Copies of Selected Letters, 1878–1894, AT&T Archives, box 2047 (hereafter cited as Hill Collection), vol. 7.
28. Knight to the American Bell Telephone Company, 15 December 1883, AT&T Archives, box 1211; Haskins to Vail, 3 April 1882, ibid.; H. B. Stone to C. J. French, 15 October 1891, 9, Miscellaneous French Letter Books, AT&T Archives.
29. Knight to the American Bell Telephone Company, 23 April 1882, Hill Collection. Reporting on the rather dismal record of neglect posted by the management of the St. Joseph, Missouri, exchange, traveling agent Knight noted: "The connections over the ex-ter line to Atchison number, I was told, from 7 to 10 a day. The opr. however said that there would be many more if the line was kept up better. He reported it down frequently, and . . . the manager of Atchison, says that the trouble is generally at St. Joe and instanced one time when the line had not worked for a couple of days and having reported that his portion of it was alright he followed it on foot to St. Jo. 25 miles and found the trouble in that City proof conclusive that the St. Jo. man was not attending to the business."
30. Henry Howard to Forbes, 8 December 1879, ibid., box 1013; Directors Minutes of the Interstate Telephone Company of New York, 5 December 1879, Southern New England Telephone Company Records Center, 227 Church Street, New Haven, Conn. (hereafter cited as SNET Archives); Walsh, *Connecticut Pioneers*, 112.

31. Walsh, *Connecticut Pioneers,* 112–13; Directors Minutes of the Inter-state Telephone Company, 27 July 1880, SNET Archives.
32. Directors Minutes of the Interstate Telephone Company, 13 July, 30 November 1881, SNET Archives.
33. Report of the Interstate Telephone Company to the Superintendent of the Census, 30 June 1881, 3, ibid.
34. C. N. Barney to the American Bell Telephone Company, 19 January 1881, AT&T Archives, box 1013.
35. *Report of the Directors of the American Bell Telephone Company to the Stockholders* (hereafter cited as *Annual Report*), 25 March 1884; 31 March 1885, 5, AT&T Archives.
36. Ibid., 25 March 1884, 23.
37. Forbes to A. D. Bullock, 8 March 1881, Presidents' Letter Books, 2A:352–59. This letter, written in response to proposals before the Illinois legislature and under consideration in Ohio, proposals that would have prohibited the consolidation of telephone companies, aptly captures Forbes's sentiments regarding the need to reorganize Bell agencies. After recounting his belief that it was "a mistake and against the interests of the public to pass such measures," Forbes assured Bullock that

> consolidation will only take place so far as groups of exchanges can be operated more cheaply and better from some central town. There is too much detail to make it an object for one company to occupy a large number of towns not closely connected in business—but where such connections naturally exist and it is desirable to give telephone connection between the exchanges each group can often be more effectively handled by one Company than by a number, each perhaps preferring a different plan to that of its neighbor. Moreover, in our test with the Western Union Company, it was demonstrated that two Telephone systems are not wanted in the same locality. Every person wants connection with all stations and unless a duplicate system is run into every station the subscribers are not served. Two Companies dividing the stations in competition are only one half the convenience to the public that one company reaching the whole will be. Everywhere people asked our licensees to settle the dispute, to get together and place all stations under one system and management as the only way to get complete facilities with the least public inconvenience from wires and fixtures.

38. Pier, *Forbes,* 108–10.
39. For a list of early agencies and exchanges that later combined with these new, regional systems see the folder entitled "Associated Companies—Boundaries of—1913," AT&T Archives, box 1010.

40. Ibid.
41. Ibid. These enterprises included the Central Telephone Company, the Midland properties, Western Telephone, the C. L. Clapp Agency, the Denville Telephone Company, the Greenburg Telephone Company, Iowa Telephone, the Reiser Agency, and the Gleason properties.
42. Ibid.
43. *Annual Report*, 30 March 1886, 2, AT&T Archives.
44. *Annual Report*, 25 March 1884, 1.
45. FCC, *Telephone Investigation*, 29; Smith, *Anatomy of a Business Strategy*, chap. 4.
46. *Annual Report*, 29 March 1882, 16.
47. Federal Communications Commission, *Report on AT&T License Contract Relations: Origins and Development of the License Contract* (Washington, D.C.: U.S. Government Printing Office, 1936), 10–11, 15–23. The Bell Telephone Company of Missouri holds the distinction of having been the first licensee to receive an "amended" perpetual agreement under the new policy in January 1881. Few perpetual agreements were awarded before 1882, however (*Annual Report*, 29 March 1882, 16–17).
48. FCC, *Telephone Investigation*, 21–23.
49. Said one Bell official in testimony later, "All of our short term contracts, especially the later ones, gave us the right at the expiration of the contracts to come in and take the property and continue the business ourselves. The Bell Company was very seriously considering [this] but instead of doing it that way, [it] sold either to the particular local company that was operating the business or to an aggregation of such companies, the right to do it, and gave them a permanent contract taking for that right . . . a certain percentage of their future profits represented by stock" (testimony of Henry L. Storke, *Western Union Telegraph Company v. American Bell Telephone Company*, exhibit C, "Evidence for the Defendant," 2 vols. [Boston, 1909], 1:180–81).
50. Almost universally the licensees had argued that the five-year contracts of late 1879 and 1880 did not furnish them with a secure arrangement under which they could make long-range plans for developing their exchange business. Such sentiments were acknowledged by O. E. Madden at the time of the fall 1880 Exchange Association meeting (*Exchange Association Report*, September 1880, 85–92).
51. Vail to Watson, 5 February 1881, GMLB 76:335–40. The same letter from Vail, outlining the expanding mission of the Electrical and

Patent Department, was sent to T. D. Lockwood, W. W. Jacques, and E. Berliner.

52. Madden to R. J. Boyd, 1 December 1881, ibid., 106:144–45; Madden to Knight, 30 March 1881, ibid., 81:315–19.

53. In April 1881 Thomas Lockwood, who, as part of Watson's Electrical Department, was in charge of assembling the company's library on patents and evaluating the devices of independent inventors, wrote Vail that "it is a fact that no electrical expert of this company is now on the road or visiting exchanges. . . . Aside from the fact that the numerous small exchanges operating under the license of the company would be materially benefited by the periodic visits of a practical man, is it not likely that the company would be indirectly benefited by the visits of such a man to the larger exchanges which would lead to careful comparisons of the present systems and of any novelties, which from time to time would arise." Subsequently, T. B. Doolittle was granted a full-time assignment in the field as one of only two special agents who, according to Lockwood, displayed any "ability of a technical character" (Lockwood to Vail, 11 April 1881, and Doolittle to the American Bell Telephone Company, 6 May 1881, both in Hill Collection, vol. 7).

54. *Annual Report*, 25 March 1884, 10.

55. Ibid.

56. The total headquarters staff numbered between sixty and sixty-seven employees in December 1885. These imprecise figures reflect the difficulty in determining the make-up of the company's permanent legal staff, which at the time comprised both outside counsel and corporate counsel. Moreover, these figures may overstate the company's employment rosters, owing to a small number of unrecorded retirements and personnel reassignments. The estimates do not include members of the Executive Committee or major officers of the corporation (figures compiled from folder entitled "American Bell Telephone Company and Predecessors—Officers and Employees—1877–1902," AT&T Archives, box 1008).

57. Those not reporting directly to the general manager were engaged in what later came to be known as corporate staff activities, such as handling the legal and financial affairs of the company. Thus Vail, as general manager of American Bell, functioned as its chief operating officer.

58. Robert S. Boyd to Vail, 18 January 1884, ibid., box 1080.

59. John Hudson to Special Agents, 1 December 1885, ibid., box 1231.

Chapter 6   The American Telephone and Telegraph Company

1. *Annual Report of the Southern New England Telephone Company,* 11 May 1885, cited in Walsh, *Connecticut Pioneers,* 145–46.
2. Vail to Hall, 16 February 1885, AT&T Archives, box 1010. On Hall's career in telephony see folder entitled "Hall, E. J.— Biography," ibid., box 1120.
3. Vail to Hall, 16 February 1885, ibid., box 1010.
4. Hall to Vail, 12 May 1885, ibid.
5. Ibid., 9.
6. Ibid., 6.
7. Vail to Joseph P. Davis, 26 May 1884; Vail to C. P. Bowditch, 28, 31 May 1884; Madden to Davis, 14 January 1885; L. F. Beckwith to Forbes, 25 February 1885; Green to Forbes, 26 February 1885, all in ibid., box 1013.
8. Beckwith to Forbes, 25 February 1885; and Hall to Vail, 3 May 1886, both in ibid., box 1101; Madden to the Executive Committee of the American Bell Telephone Company, 7 January 1885, and Doolittle to Hall, 7 March 1885, both in ibid., box 1013.
9. Madden to the Executive Committee of the American Bell Telephone Company, 7 January 1885, ibid., box 1013.
10. *New York Commercial News,* 18 December 1886.
11. *Philadelphia News,* 7 December 1886.
12. Hall to Vail, 3 May, 29 September 1886, AT&T Archives, box 1218.
13. Hall to Hudson, 21 January 1888, ibid., box 1011.
14. Hall to Hudson, 20 November 1886; and 6 January 1887, ibid., box 1218.
15. Hall to Vail, 29 September 1886, ibid.
16. Hall to Hudson, 21 January 1888, ibid., box 1011.
17. "Telephone Revenues, Expenses, Net Telephone Earnings and Average Plant in Service, 1888–1925," AT&T Archives (uncatalogued).
18. George Bartlett Prescott, *The Electric Telephone,* 2d ed. (New York: D. Appleton & Company, 1890), 783.
19. Frank B. Jewett, "The Telephone Switchboard: Fifty Years of History," in *The Telephone: An Historical Anthology,* ed. George Shiers (New York: Arno Press, 1977).
20. It was not until the end of 1890 that the American Bell Company reported to shareholders that "a substantial and gratifying beginning [had] been made in metallic circuit service in the exchanges." The first time that figures on the number of stations connected to central exchanges through upgraded metallic circuits (11,584) were reported was in American Bell's *Annual Report* for 1892; the number had more than doubled by the end of the year (*Annual Report,* 21 March 1891,

6; 28 March 1893, 5, 8–9, AT&T Archives).

21. The details of the 1886 plan for consolidation are discussed at length in Hall to Bowditch, 22 February 1886, AT&T Archives, box 1010.
22. Ibid., 2.
23. Ibid., 2–3.
24. Ibid., 9.
25. Ibid.
26. William Kerr to Hubbard, 25 October 1885, ibid., box 1115; Hubbard to Forbes, 12 November 1885, ibid.; Forbes to Hubbard, 19 November 1885, Presidents' Letter Books, 3G:253; Hubbard to Forbes, 29 October 1885, AT&T Archives, box 1115.
27. Forbes to Hubbard, 9 November 1885, Presidents' Letter Books, 3G:245.
28. Forbes to S. F. Emmons, 6 March 1886, ibid., 432.
29. Hubbard to Stockton, 7 April 1888, AT&T Archives, box 1115.
30. Ibid.
31. Stockton to Hubbard, 3 May 1888, Presidents' Letter Books, 4G:201.
32. Francis Blake to Stockton, 7 May 1888, AT&T Archives, box 1115.
33. Hall to Hudson, 9 January 1889, ibid., box 1010.
34. Ibid.
35. A. S. Hibbard, J. J. Carty, and F. A. Pickernell, "The New Era in Telephony," *Eleventh Meeting of the National Telephone Exchange Association, September 9–11, 1889* (Brooklyn: Press of the Brooklyn Eagle Book Printing Department, 1889), 34.

Chapter 7 The Hudson Years

1. Danielian has described Hudson as "unbending and unimaginative," an autocrat who confused aristocracy with tyranny," and "the hired man of the Bostonians controlling the board" (Danielian, *AT&T*, 58, 71).
2. Ibid., 97–98.
3. Edward J. Hall, Jr., "Corporate Organization," in *Twelfth Meeting of the National Telephone Exchange Association, September 9th and 10th, 1890* (Brooklyn: Press of the Brooklyn Eagle Book Printing Department, 1890), 43–49, 55.
4. Ibid., 51.
5. ibid., 51, 53–54.
6. Ibid., 57, 51.
7. "Report of Proceedings at a Meeting of the Standing Committee on Telephone Cables, April 14th and 15th, 1891," AT&T Archives.
8. Minutes of the Switchboard and Telephone Apparatus Committee, 21–23 July 1891, 8, 99, 124, ibid.

9. Statistics concerning the capital investment and earnings of the Long Distance Company are available in a folder entitled "Long Lines—Review of Earnings—1888–1923," ibid., box 1251.
10. "Report of Proceedings at a Meeting of the Standing Committee on Telephone Cables, April 14th and 15th, 1891," 64–65.
11. *Annual Report,* 28 March 1893, 5, 9.
12. *Annual Report,* 27 March 1894, 9.
13. "Early Development of Telephone Accounts: Bell Telephone Association, July 1, 1877–July 1, 1878," AT&T Archives, box 1001; Thomas Sherwin, "Subject Accounts," 25 April 1884, ibid., box 1007.
14. Thomas Sherwin, "Subject: Distribution of Expenses," 30 December 1887, ibid., box 1007.
15. Hall to Hudson, 19 November 1890, ibid., box 1001.
16. Thomas Sherwin, "Subject Accounts," 1 January 1891, ibid.
17. Ibid.
18. R. F. Fiske to Thomas Sherwin, 6 January 1894, and S. A. Richardson to Thomas Sherwin, 15 October 1894, both in ibid., box 1007.
19. *Annual Report,* 28 March 1893, 8–9, 12.
20. FCC, *Telephone Investigation,* 130.
21. See Chapter 8 for an account of how service and operating standards suffered in the Central Union's territory owing to the pressures of competition.
22. Ibid., 52; *Annual Report,* 27 March 1900, 8–9, AT&T Archives. In 1899 "new construction" expenditures of the licensees included the following allocations: exchange construction, $15.9 million; toll-line construction, $8.1 million; real estate investment, $2.1 million.
23. FCC *Telephone Investigation,* 21, 45.
24. *Annual Report,* 27 March 1894, 14.
25. Stehman, *Financial History,* 60–61.
26. Ibid., 60–62.
27. Ibid., 61–62. Stehman has compiled the following figures regarding Bell's equity transactions for the period 1879–97.

Company	Cash Paid in for Stock	Stock Issued
New England Company and Bell Company	$ 100,000.00	$ 650,000
National Bell sale of stock at varying prices	430,000.00	200,000

Company	Cash Paid in for Stock	Stock Issued
American Bell bonus shares to National Bell stockholders and sale of stock at par	14,900,000.00	19,150,000
American Bell sale of stock at the market price		
November 1894		
Stockholders at 190	310,080.00	163,200
Auction at 189.70	638,909.60	336,800
July 1895		
Stockholders at 194	805,294.00	415,100
Auction at 196	1,146,404.00	584,900
June 1896		
Stockholders at 200	3,449,000.00	1,724,500
Auction at 203.50	865,892.50	425,500
April 1897		
Stockholders at 214	4,785,682.00	2,236,300
Total	$27,431,262.10	$25,886,300

28. Owing to the company's secure financial posture, the Hudson administration could negotiate a 4 percent rate on its bonded debt. Since the yield on American Bell's common stock, valued at par, was 15 percent, debt financing offered management a much more favorable source of low-cost funding. Even if the market value of American Bell equity is taken into account—on the Boston Stock exchange Bell shares sold at between $273.00 and $386.00 for the year 1899—the effective cost of equity was still usually greater (5.5 percent and 3.9 percent, respectively) than the claim against earnings registered by an equivalent amount of debt. The last point has been made by Lawrence Steadman in research that he conducted into American Bell and AT&T financial practices (Cambridge Research Institute, "Fund Flows in the American Bell Telephone Company and the American Telephone and Telegraph Company, 1881–1935: Preliminary Commentary and Analysis" [Cambridge, Mass., 1979]).

29. FCC, *Telephone Investigation*, 430–37.

30. "Consolidations," 4 January 1896, AT&T Archives, box 1382. This memo is unsigned; however, there is good cause for believing that E. J. Hall, Jr., was its author. Many of the arguments that Hall had

made earlier in other correspondence with Hudson are repeated in this document.

31. Hall to Hudson, 25 November 1896, ibid.
32. Ibid.
33. FCC, *Telephone Investigation,* 12–14.
34. The figures compiled from American Bell's annual reports for the period 1894–96 are as follows: 1894—assets $45.6 million, net income $3.1 million; 1895—assets $51.3 million, net income $3.2 million; 1896—assets $55.3 million, net income $3.4 million.
35. FCC, *Telephone Investigation,* 135. See fig. 2.

Chapter 8   Interregnum

1. Biographical material on Frederick P. Fish can be found in a folder entitled, "Fish, Frederick P.—Biography—1850–1930," AT&T Archives, box 1006.
2. Quote from "Proceedings of the Bar Association of the City of Boston and of the District Court of the United States for the District of Massachusetts, in Memory of Frederick Perry Fish," 26 December 1931, ibid.
3. Ibid., 6; "Remarks of William K. Richardson in Memory of Frederick P. Fish," 26 December 1931, 9, ibid.
4. "Remarks of Robert P. Clapp in Memory of Frederick P. Fish," 26 December 1931, 12, ibid.
5. Danielian, *AT&T,* 57–66.
6. Thomas P. Hughes, *Networks of Power: Electrification in Western Society, 1880–1930* (Baltimore: Johns Hopkins University Press, 1983), 15. Hughes notes that as a system grows it builds momentum and that this arises from the "involvement of persons whose professional skills are particularly applicable to the system." The various organizations that together make up the system are described by Hughes as the system's culture. In the case of the Bell System, the principal organizational sources of momentum were the technical departments of AT&T and the management of its long-distance enterprise, both of which were deeply involved in activities of network development and integration.
7. Davis to Fish, 23 October 1902, AT&T Archives, box 1360.
8. Sherwin, "Subject Accounts," 1 January 1903, ibid.
9. Haskins and Sells to Fish, 20 June 1903, 9, 2–3, ibid. Haskins and Sells stated that "the separation of expenses for exchange and toll service, as contemplated in your circular, is questionable from an

accounting standpoint. Necessarily, the segregation is too frequently on some arbitrary basis, therefore, not creating dependence upon the reliability of the results. Perhaps our position on this point is best sustained by the railway companies and the Interstate Commerce Commission. For many years they attempted to separate the expenses as between freight and passenger traffic, but finally abandoned the effort through becoming convinced of the unreliability of the results which should be arrived at." The auditor of the Chicago Telephone Company apparently shared these very views (see Arthur D. Wheeler to Sherwin, 24 July 1903, ibid., box 1360).

10. E. B. Field to Sherwin, 19 August 1903, ibid., box 1320; see also Edmund W. Longley to Sherwin, 6 August 1903, ibid.

11. A. R. Schellenberger to Sherwin, 18 July 1903; ibid.; the quote is from pp. 1–2.

12. Walter B. Butler to Sherwin, 28 July 1903, ibid.

13. Sherwin to Fish, 11 September 1903, 1–2, ibid.

14. Ibid., 2.

15. Ibid., 15–16.

16. American Bell Telephone Company, "Subject—Accounts," 1 January 1904, ibid., box 1360. The circular was issued by the American Bell Company rather than by the new parent, AT&T, because until the former was dissolved in the 1920s, many of the license contracts (royalties and patents) were still in its custody. This might have been owing to certain debt covenants that prevented American Bell from divesting or transferring certain assets that would have adversely affected its earnings potential.

17. Charles G. DuBois, *A Brief History of Telephone Accounting* (New York: AT&T, 1913), 7, AT&T Accounting Library, 550 Madison Ave., New York, N.Y.

18. Danielian, *AT&T*, 58. Danielian quotes Theodore Vail as writing Senator Crane shortly before both men were elected to AT&T's board:

> The worst of the opposition has come from the lack of facilities afforded by our companies, that is, either no service, or poor service. For this circumstances beyond our control are to a great extent responsible, as it was in the early days, very difficult to provide money.
>
> To meet these increasing demands, increasing amounts of money will be needed each year. A low estimate for the next five years would be $200,000,000—every probability points to a larger sum.
>
> These demands necessitate a broad financial policy covering a period of no less than five years.

19. C. J. French, "Subject: New Work for 1901," AT&T Archives, box 1377.

20. Danielian, *AT&T*, 58.

21. *1902 Annual Report*, 6, AT&T Archives; Stehman, *Financial History*, 109–10; Federal Communications Commission, *Proposed Report: Telephone Investigation* (Washington, D.C.: U.S. Government Printing Office, 1938).

22. *1905 Annual Report*, 6, 16; *1906 Annual Report*, 22, AT&T Archives.

23. *1906 Annual Report*, 22.

24. Figures were assembled from AT&T's *Annual Reports* for the years 1900 and 1906.

25. Stehman, *Financial History*, 108.

26. Angus S. Hibbard to Hall, 12 January 1891, AT&T Archives, box 2021.

27. J. J. Carty to H. B. Thayer, 1 March 1905, 5, ibid., box 1357.

28. Davis to Fish, 15 March 1902, ibid., box 1341.

29. Fish to P. Yensen, 9 April 1902, ibid.

30. T. Spencer to Fish, 2 April 1903, and C. H. Wilson to Davis, 19 May 1903, both in ibid.

31. Fish to John Sabin, 14 September 1903, Presidents' Letter Books, 3G:291–94.

32. Davis to Fish, 14 January 1904, AT&T Archives, box 2045. Those listed as entering into an "engineering affiliation" with AT&T's department included: Bell Telephone of Canada, the Central District and Printing Telegraph Company, the Central Union Telephone Company, the Chicago Telephone Company, the Cleveland Telephone Company, the Colorado Telephone Company, the Michigan Telephone Company, the Missouri and Kansas Company, the Nebraska Telephone Company, the Northwestern Telephone Exchange Company, the Pacific States Telephone and Telegraph Company, the Pennsylvania Telephone Company, the Southwestern Telegraph and Telephone Company, and the Wisconsin Telephone Company.

33. Davis to Fish, 23 March 1903, ibid., box 1341.

34. FCC, *Telephone Investigation*, 30–31.

35. Davis to Fish, 11 February 1903, AT&T Archives, box 1341.

36. "Engineers Department—Annual Report," 31 December 1906, 2, ibid., box 2045.

37. FCC, *Telephone Investigation*, 30–31.

38. Transcript of Fish to Central Union Telephone Company, 21 November 1902, attached to "Resolution of Special Meeting of Directors of the Central Union Telephone Company," 9 January 1903, AT&T Archives, box 1352.

39. FCC, *Telephone Investigation,* exhibit 1060B, 395.
40. For a sampling of the various opinions among the AT&T management concerning this issue see George V. Leverett to Fish, 17 October 1901, AT&T Archives, box 1355; Hall to Fish, 22 October 1901, ibid., box 1375; Davis to Fish, 23 October 1901, ibid.; and C. E. Yost to Fish, 12 September 1902, ibid., box 1353.
41. Fish to Wheeler, 20 July 1906, Fish Letter Book, 5:163, AT&T Archives. Similar letters urging a "radical change in policy in the matter of spending money" and of the "vital importance [of] hold[ing] up construction to the last dollar" were sent to all Bell Company presidents and to officers of AT&T Long Lines.

Chapter 9   The Vail Years

1. Danielian, *AT&T,* 57–69.
2. *1907 Annual Report,* 17, 18, AT&T Archives.
3. Ibid., 18–20.
4. By the end of 1907 the following states had established public agencies with regulatory authority over the telephone business: Indiana (1885); Mississippi (1892); Louisiana (1898); South Carolina (1904); and Nebraska, Nevada, Oklahoma, and Alabama (all 1907). Between 1908 and the end of 1911, fourteen additional states vested public-utility authorities with specific telephone regulatory powers: Vermont, Washington, Maryland, New Jersey, New York, Oregon, Michigan, New Hampshire, New Mexico, North Dakota, Ohio, California, Connecticut, and Kansas. Twelve more states added telephone regulation to the more traditional duties of their utility authorities over the next two years.
5. Martin J. Schiesl, *The Politics of Efficiency: Municipal Administration and Reform in America, 1880–1920* (Berkeley: University of California Press, 1977). For a contemporary account of this movement, see "Municipal Ownership versus Adequate Regulation," chap. 2 in Clyde Lyndon King, ed., *The Regulation of Municipal Utilities,* National Municipal League Series (New York: D. Appleton & Co., 1918).
6. FCC, *Telephone Investigation,* 134, 129.
7. Fish to Sherwin; to Charles F. Cutler; to Charles H. Wilson; and to W. T. Gentry—all 20 July 1906, Fish Letter Book, 5:157, 158, 159, and 160, respectively. In the letter to Gentry, vice president and general manager of the Southern Bell Telephone and Telegraph Company, Fish emphasized "the necessity of reducing expenditures

and curtailing construction." He continued: "The high price of labor and material and the fact that you cannot get supplies in a satisfactory way, are quite sufficient reasons for this to the public and your organization; but confidentially I say to you there are other reasons that are important." Letters such as this, urging an immediate retrenchment in capital spending, were sent to all Bell Company officers.

8. The scale of the Vail retrenchment is evident in the figures for telephone plant expansion compiled by the FCC during this period (see app. B).

9. Danielian, *AT&T,* 58.

10. "Provisional Estimate by the Cincinnati and Suburban Bell Telephone Company of the Plant Requirements for the Year 1909," AT&T Archives, box 1348.

11. Fish appears to have channeled an enormous proportion of the available resources into areas subject to intense competition, even though the chance of earning a respectable return on such investment was clearly minimal, given the existing market conditions. Vail changed the criteria upon which decisions on the allocation of capital were based.

12. See form 528A, dated 30 September 1908, in ibid.

13. E. S. Bloom, "The Provisional Estimate," in Thayer to Nathan C. Kingsbury, 16 April 1912, ibid., box 21.

14. FCC, *Telephone Investigation,* 507, 519.

15. Ibid., 519.

16. Vail to Senator Crane, [c. 1901], 6–7, AT&T Archives, box 1080.

17. Minutes of the AT&T Company Executive Committee, 25 January 1907, ibid., box 1377.

18. Fish to John I. Waterbury, 3 April 1907, Fish Letter Book, 6:6; Fish to Vail, 4 April 1907, ibid., 7. Fish wrote Waterbury that he thought "it would be wise if we could get a couple of copies of the By-Laws of The Steel Company and possibly of some of the big railway companies and let Mr. Vail and myself see how the work is distributed in some of the larger corporations."

19. Organization Charts of the AT&T Company, 1905–1957, AT&T Secretary's Office, 550 Madison Ave., New York, N.Y.

20. *Application of Some General Principles of Organization* (New York: AT&T, 1909), 4, AT&T Archives, box 2029.

21. Central office equipment design in this case involved specifying the capacity and other design characteristics required by current and forecasted service demand. This function bore little or no relation to

the kind of design activity carried out by the development and manufacturing arms of the business.

22. Wilson to Carty, 24 March 1908; Wilson, "Change in Organization," 24 March 1908; and Wilson to H. S. Brooks, 24 March 1908, all ibid.

23. Carty to Hall, 17 July 1907, 3, ibid., box 6.

24. That link had been established during the early 1880s through provisions in the permanent license contracts obligating Western Electric to supply local Bell companies with all the equipment they required for their operations.

25. Carty to Hall, ibid., 4–5.

26. Hall to Vail, 27 September 1909, AT&T Archives, box 1010.

27. FCC, *Telephone Investigation*, 20–21.

28. Field to Carty, 8 September 1909, AT&T Archives, box 2029.

29. Ibid., 4–5; emphasis in the original.

30. Hall to Vail, 27 September 1909, ibid., box 1010.

31. AT&T, *Application of Some General Principles of Organization*, 2, October 1909, ibid., box 2029.

32. H. F. Thurber to Carty, 16 May 1908, AT&T Archives, box 2029.

33. See folder entitled "Central Union Telephone Company— Organization and Development—1883–1908," ibid., box 1039.

34. In 1909, B. E. Sunny, who had been in charge of Central Union operations before assuming a position at AT&T headquarters, wrote Vail: "There can be no doubt that a strong central organization, to handle the service of these five important states, with a well worked policy applicable to all stations, would make for an increase in efficiency, and result in a savings of from $250,000 to $500,000 per annum" (B. E. Sunny to Vail, 26 April 1909, ibid.).

35. L. G. Richardson to Sunny, 31 May 1912, ibid., box 1040.

36. E. S. Bloom to Thayer, 26 July and 26 December 1919, ibid., box 43.

37. Bloom to Thayer, 14 February 1920, ibid.

38. Minutes of Special Meeting of the Board of Directors of the Cleveland Telephone Company, 29 January 1920, 2, AT&T Regulatory Project, Ohio Bell submission, AT&T Archives.

39. See: R. W. Garnet, "The Chesapeake and Potomac System," 13 February 1979, AT&T Archives. When in 1896 Congress established mandatory telephone rates for the District of Columbia, and the Chesapeake and Potomac Telephone Company only partially complied, numerous suits were filed by subscribers against the firm. Among these *J. Forest Manning v. The Chesapeake and Potomac Company* became a cause célèbre, spawning a thorough examination of C&P's allegations that the new metallic-circuit service could not

be profitably provided at the lawful rate. During the court's investigation, the consolidated nature of C&P's capital accounts (both
Bell's Maryland and D.C. telephone properties were owned and
administered by the Chesapeake and Potomac Company of New
York, a holding company) complicated the assignment of such costs
to particular regulatory jurisdictions. In its final opinion, the court
rejected C&P's estimates for D.C. capital expenditures, noting that
"the evidence does not show that forty percent of the bond sales
(negotiated on a consolidated basis) was actually expended in better-
ments in the District, nor does it show that forty percent of general
expenses is the real and just proportion to be born by the Washington
plant."

40. "Memorandum as to a New Jersey Corporation," 20 October 1921,
ibid., box 5. See also Cutler to Fish, 16 January and 6 February 1905,
ibid., box 1012; and French to Fish, 8 May 1905, 11, Miscellaneous
French Letter Books.

41. The statewide incorporation of New Jersey Bell was postponed until
1929 owing to a 1909 mortgage covenant prohibiting New York
Telephone from divesting its northern New Jersey properties.

42. See folder entitled "Southwestern Group of Associated Companies—
Organization—1910–1912," ibid., box 77; see also "Southwestern
Telephone and Telegraph Company History by G. W. Foster,
1880–1915," ibid., box 1042.

43. F. S. Pillsbury to Sabin, 15 February 1900, ibid., box 1369.

44. Report of the Board of Public Utilities for Los Angeles, 27 May 1910,
ibid.

45. In August 1915, California's Railroad Commission assumed jurisdiction over telephone rates in the city of Los Angeles. This event,
coupled with a $400,000 shortfall in city revenues the next year, in
effect dissipated support for home-rule legislation.

46. One of the more notable instances of this kind of opposition came
from the traffic engineers at the New York Telephone Company.
After learning that their company's northern New Jersey properties
were about to be spun off as part of a plan involving the formation of a
separate New Jersey company, they argued that "it would be possible
to handle traffic between New York City and New Jersey offices if the
New Jersey area were operated by a separate company, but it [could] be
handled better by a single Company." Pointing out that "the Metropolitan area is distinctly a unit from an operating standpoint," the
engineers concluded that the structural separation of New York and
New Jersey properties would engender an "unnatural division" of

labor (Engineering Department, New York Telephone Company, "New Jersey as a Separate Company from a Traffic Standpoint," ibid., box 5).

47. Green to Forbes, 12 December 1879 and 20 April 1880, ibid., box 1004.

48. "Subject: Consolidation—New York Telephone Companies," 20 October 1899, ibid., box 1379.

49. Doolittle to Hudson, 1 June 1895, ibid., box 1212.

50. Vail to George J. Gould, 25 March 1908, ibid., box 47. See also Vail to Robert Winsor, 26 March 1909, in which Vail claimed that it was "absolutely necessary for the protection of New York [City] property to consolidate it with the entire State property," since "any investigation of its earnings would lead to more or less attack" (ibid.).

51. The number of directly competing independent exchanges peaked in 1911. That year Bell and the independent exchanges competed at 1,384 points, and independents competed with other independents enjoying Bell connections at 906 points. By 1 October 1913 these figures had dropped to 1,153 and 618, respectively (see app. A).

52. George Wickersham to J. A. Fowler, 29 August 1912, "Section 3," Box 38, Record Group 60, National Archives, Washington, D.C.

53. Wickersham to Charles A. Prouty, 7 January 1913, 6, "Section 4," Box 36, ibid.

54. Independent Telephone Association of America to G. Carroll Todd, 18 December 1913, "Section 6," box 38, ibid.

55. Kingsbury to James McReynolds, 19 December 1913, "Section 8," ibid.

Chapter 10 Reflections

1. In his *Networks of Power*, Thomas P. Hughes discusses how the professionalization of a small but expanding group of electrical engineers, the development of professional engineering organizations and publications, and the founding and evolution of electrical engineering research laboratories provided a broad institutional structure for sustaining the momentum of the evolving electrical supply systems of the 1890s and the early twentieth century in the United States and Germany. Within the Bell System, a similar systems momentum was evident, this time fostered by the increasingly professional management of AT&T and by the proliferation of both formal and informal (the Meetings of the Cable and Switchboard committees, for instance) organizations devoted to solving the technical problems

associated with the expansion and operation of the national network. Together these units provided the institutional framework within which momentum was generated (see Hughes, *Networks of Power,* 140–74).

2. See, for example, Chandler, *Strategy and Structure* and *The Visible Hand.*

Bibliography ❧

Most of the material used in this book can be found in the American Telephone and Telegraph Company Archives, 195 Broadway, New York, N.Y. The AT&T Archives contain an extensive collection of material documenting the early corporate development of AT&T and its associated Bell System companies.

Among the most useful sources of information is the company's letter book collection, which includes about six hundred volumes of general manager correspondence covering the period October 1877 to July 1930. Despite a gap between the crucial years 1901 and 1907, the General Managers' Letter Books provide a broad view of company operations and organization. Corporate policy development, on the other hand, can be gleaned from the forty-three volumes of Presidents' Letter Books, which cover basically the same period. In addition, there are the private letter books of Theodore N. Vail; E. J. Hall, Jr.; George Bradley; and C. J. French. Together these letter books represent a substantial portion of the outgoing correspondence of the company's officers and management during its formative stages.

Incoming correspondence, including but not limited to special agent reports; licensee and operating telephone company reports; monthly and annual reports of various departments; financial correspondence and records; and documents dealing with the acquisition of independent telephone companies, state and federal government relations, and organization plans can be found in the five hundred boxes of loose material catalogued and indexed in the archives. For the most part, this material covers the years 1877 to 1930. Material of more recent vintage is less rich in content.

The archives also house a complete collection of annual reports for AT&T and its predecessor enterprises. An incomplete collection of annual reports submitted to AT&T by its associated companies is also available.

Supplementing this material are Exchange Association Conference minutes, minutes of the meetings of the Switchboard and Cable committees and the Telephone Rate Conference reports. These represent an invaluable source of information both on the specific subjects covered and on the dialogue that occurred between Bell Company officials and their licensees. The last three are particularly useful for understanding the more arcane technical and economic problems of the business during the period of network formation.

The AT&T Archives also contain an extensive collection of books on communications in general and the Bell System in particular. The following list of books comprises those books that I found most useful.

Bell Telephone Laboratories. *Engineering and Operations in the Bell System*. Murray Hill, N.J., 1971.

Bode, H. W. *Synergy: Technical Integration and Technological Innovation in the Bell System*. Murray Hill, N.J.: Bell Telephone Laboratories, 1971.

Bray, Douglas W.; Campbell, Richard J.; and Grant, Donald L. *Formative Years in Business: A Long Term AT&T Study of Managerial Lives*. New York: Wiley-Intrascience, 1974.

Brock, Gerald W. *The Telecommunications Industry: The Dynamics of Market Structure*. Cambridge, Mass.: Harvard University Press, 1981.

Bruce, Robert V. *Bell: Alexander Graham Bell and the Conquest of Solitude*. Boston and Toronto: Little, Brown & Co., 1973.

Casson, Herbert N. *The History of the Telephone*. 9th ed. Chicago: A. C. McLurg, 1917.

Chandler, Alfred D., Jr. *Strategy and Structure: Chapters in the History of the American Industrial Enterprise*. Cambridge, Mass.: MIT Press, 1962.

———. *The Visible Hand: The Managerial Revolution in American Business*. Cambridge, Mass.: Harvard University Press, 1977.

Conway, Connie Jean. "Theodore Vail's Public Relations Philosophy." *Bell Telephone Magazine* 37, pt. 1 (Autumn 1958), 39–46; pt. 2 (Winter 1958/59), 522–59.

Coon, Horace. *American Tel & Tel: The Story of a Great Monopoly.* New York and Toronto: Longmans, Green & Co., 1939. Reprint. Freeport, N.Y.: Books for Libraries Press, 1971.

Danielian, N. R. *AT&T: The Story of Industrial Conquest.* New York: Vanguard Press, 1939.

Fagen, M. D., ed. *A History of Engineering and Science in the Bell System: The Early Years, 1875–1925.* N.p.: Bell Telephone Laboratories, 1975.

Federal Communications Commission. *Proposed Report: Telephone Investigation.* Washington, D.C.: U.S. Government Printing Office, 1938.

Gabel, Richard. "The Early Competitive Era in Telephone Communication, 1893–1920." *Law and Contemporary Problems* 34 (Spring 1969), 340–59.

Harlow, Alvin F. *Old Wires and New Waves: The History of the Telegraph, Telephone and Wireless.* 1936. Reprint. New York: Arno Press, 1971.

Hounshell, David A. "Elisha Gray and the Telephone: On the Disadvantages of Being an Expert." *Technology and Culture* 16 (April 1975), 133–61.

Hughes, Thomas P. *Networks of Power: Electrification in Western Society, 1880–1930.* Baltimore: Johns Hopkins University Press, 1983.

Lockwood, Thomas D. *Practical Information for Telephonists.* New York: W. J. Johnston, 1882.

Miller, Kempster B. *American Telephone Practice.* 3d ed. New York: McGraw Publishing Co., 1900.

Prescott, George Bartlett. *The Electric Telephone.* 2d ed. New York: D. Appleton & Co., 1890.

Reid, James D. *The Telegraph in America: Its Founders, Promoters and Noted Men.* New York: John Polhemus, Publisher, 1886. Reprint. New York: Arno Press, 1974.

Rhodes, Frederick Leland. *Beginnings of Telephony.* New York: Harper & Brothers, 1929.

Sobel, Robert. *The Entrepreneurs: Explorations within the American Business Tradition.* New York: Waybright & Talley, 1974.

Stehman, J. Warren. *The Financial History of the American Telephone and Telegraph Company.* Boston: Houghton Mifflin Co., 1925.

Tosiello, Rosario J. "The Birth and Early Years of the Bell Telephone System, 1876–1880." Ph.D. diss., Boston University, 1971.

U.S. Congress, Senate, Temporary National Economic Committee. *Investigation of the Concentration of Economic Power: Hearings.* 76th Cong., 3d sess. Washington, D.C.: U.S. Government Printing Office, 1939–41.

Walsh, J. Leigh. *Connecticut Pioneers in Telephony: The Origin and Growth of the Telephone Industry in Connecticut.* New Haven: Telephone Pioneers of America, 1950.

Index 🖋

References to illustrations and tables are printed in italic type

Robert W. Garnet is currently engaged in a study of the recent restructuring of the Bell System. He holds a Ph.D. degree in American history.

THE JOHNS HOPKINS UNIVERSITY PRESS

The Telephone Enterprise

*This book was composed in Goudy Old Style by Brushwood Graphics Studio,
from a design by Sheila Stoneham.
It was printed on 60-lb. Warren's Olde Style paper
and bound in Holliston Roxite A cloth by
the Maple Press Company.*